Robotics
Theory and Industrial Applications

Second Edition

by

Larry T. Ross
Department Chair—Department of Technology
Eastern Kentucky University
Richmond, KY

Stephen W. Fardo
Foundation Professor—Department of Technology
Eastern Kentucky University
Richmond, KY

James W. Masterson
Professor Emeritus—Department of Technology
Eastern Kentucky University
Richmond, KY

Robert L. Towers
Retired Professor—Department of Technology
Eastern Kentucky University
Richmond, KY

Publisher
The Goodheart-Willcox Company, Inc.
Tinley Park, Illinois
www.g-w.com

Copyright © 2011
by
The Goodheart-Willcox Company, Inc.

Previous edition copyright 1996
Previously published and copyrighted as *Robotics Technology*
by The Goodheart-Willcox Company, Inc.

All rights reserved. No part of this work may be reproduced, stored,
or transmitted in any form or by any electronic or mechanical means,
including information storage and retrieval systems, without the prior
written permission of The Goodheart-Willcox Company, Inc.

Manufactured in the United States of America.

Library of Congress Catalog Card Number 2010003091

ISBN 978-1-60525-321-3

1 2 3 4 5 6 7 8 9 – 11 – 15 14 13 12 11 10

The Goodheart-Willcox Company, Inc. Brand Disclaimer: Brand names, company names, and illustrations for products and services included in this text are provided for educational purposes only and do not represent or imply endorsement or recommendation by the author or the publisher.

The Goodheart-Willcox Company, Inc. Safety Notice: The reader is expressly advised to carefully read, understand, and apply all safety precautions and warnings described in this book or that might also be indicated in undertaking the activities and exercises described herein to minimize risk of personal injury or injury to others. Common sense and good judgment should also be exercised and applied to help avoid all potential hazards. The reader should always refer to the appropriate manufacturer's technical information, directions, and recommendations; then proceed with care to follow specific equipment operating instructions. The reader should understand these notices and cautions are not exhaustive.

The publisher makes no warranty or representation whatsoever, either expressed or implied, including but not limited to equipment, procedures, and applications described or referred to herein, their quality, performance, merchantability, or fitness for a particular purpose. The publisher assumes no responsibility for any changes, errors, or omissions in this book. The publisher specifically disclaims any liability whatsoever, including any direct, indirect, incidental, consequential, special, or exemplary damages resulting, in whole or in part, from the reader's use or reliance upon the information, instructions, procedures, warnings, cautions, applications, or other matter contained in this book. The publisher assumes no responsibility for the activities of the reader.

Library of Congress Cataloging-in-Publication Data

Robotics : theory and industrial applications / by Larry T. Ross ... [et al.]. -- 2nd ed.
       p. cm.
   Previously published, under title: Robotics technology, 1996.
   Includes index.
   ISBN 978-1-60525-321-3
   1. Robotics. I. Ross, Larry T. II. Masterson, James W. Robotics technology.
   TJ211.M3672 2010
   629.8'92--dc22

2010003091

Introduction

Robotics: Theory and Industrial Applications is an introductory text that explores many aspects of robotics in a basic and easy-to-understand manner. The key concepts are discussed using a "big picture" or systems approach that greatly enhances student learning. Many applications and operational aspects of equipment and robotic systems are discussed.

We continually considered the needs of both students and instructors while preparing this comprehensive text. Several additions have been made to this edition to make this text even more comprehensive, including:

- **Updated content.** The content of the text has been revised and reorganized, with new topics added to reflect changes in the field of robotics technology. The text has continually been updated through quality checks to ensure technical accuracy.
- **Color illustrations and photos.** Numerous photos and illustrations have been updated and replaced to reflect changes in the field and in technology, in general. Additional colors have been added to illustrations for emphasis and clarity.
- **Chapter topic listing and learning objectives.** Each chapter begins with a list of the chapter content and clear performance objectives. This simple preface to the content introduces students to the organization and goals of each chapter.
- **Learning extensions.** Topic-related activities have been added to the end of each chapter to provide extended learning opportunities for students.

Organization of the Text

The development of the computer created what some experts have called the "Second Industrial Revolution." Many consider robots to be the prime movers of this revolution. The use of robots in our society has increased over the past decades, and will continue to increase. It is more important than ever to have more than a basic knowledge of robots and robotics technology.

This text is a comprehensive approach to learning the technical aspects of robotics. It is divided into four units, covering broad areas of robotic principles, power supplies and movement systems, sensing and end-of-arm tooling, and control systems.

Unit I is devoted to the basic principles of robotic technology. Chapter 1 introduces industrial robotics with a discussion of some historical events in the course of development and use of industrial robotics in the area of automation. Chapter 2 prepares a solid foundation for understanding the characteristics and fundamentals of robotics, including basic components and operation. Chapter 3 provides an overview of programming languages and techniques used to program industrial robots. Chapter 4 explores the many applications for industrial robots.

Unit II addresses robotic power supplies and movement systems. Chapter 5 provides an overview of the electromechanical systems used with robots. Alternating current and direct current systems are discussed in detail. Chapter 6 presents fluid power systems, which include hydraulic and pneumatic systems. These power sources are used for numerous robotic applications.

Unit III presents robotic sensing systems and end-of-arm tooling. Chapter 7 discusses the various sensors commonly used by robots to gain information about its external environment. Some of these include tactile sensors, proximity sensors, and photoelectric sensors. Chapter 8 provides information about various end effectors and tools used to move workpieces from one location to another within a robot's work envelope.

Unit IV covers robotic control systems, basic maintenance, and implementation planning. Chapter 9 presents the basics of digital electronics, which includes information on microcomputers and microprocessors. Chapter 10 explains how the robot controller communicates with peripheral equipment found in robotic workcells, including vision systems. Chapter 11 provides an overview of maintenance procedures. The benefits of preventive maintenance are discussed and a general plan for implementing preventive maintenance is included. Chapter 12 discusses some of the major factors to consider when using robotic systems in an industrial environment. Chapter 13, "The Future of Robotics," presents the use of robots outside the factory, artificial intelligence and expert systems, and suggested course work and training related to the field of robotics.

Appreciation

The authors would like to thank the many companies that provided photographs and technical information during preparation of the manuscript. With their cooperation, much up-to-date material was provided to make the book more valuable for the intended market.

L. Tim Ross
Stephen W. Fardo
James W. Masterson
Robert L. Towers

Department of Technology
Eastern Kentucky University
Richmond, Kentucky 40575
www.eku.edu/technology

Brief Contents

Unit I—Principles of Robotics
Chapter 1: Introduction to Industrial Robotics 11
Chapter 2: Fundamentals of Robotics 20
Chapter 3: Programming the Robot 59
Chapter 4: Industrial Applications 81

Unit II—Power Supplies and Movement Systems
Chapter 5: Electromechanical Systems 115
Chapter 6: Fluid Power Systems 139

Unit III—Sensing and End-of-Arm Tooling
Chapter 7: Sensors 171
Chapter 8: End Effectors 195

Unit IV—Control Systems and Maintenance
Chapter 9: Computer Systems and Digital Electronics 211
Chapter 10: Interfacing and Vision Systems 241
Chapter 11: Maintaining Robotic Systems 253
Chapter 12: Robots in Modern Manufacturing 265
Chapter 13: The Future of Robotics 281

Table of Contents

Unit I—Principles of Robotics

Chapter 1: Introduction to Industrial Robotics 11
- 1.1 Early Robots 12
  - Robots in Literature • The Advent of Computers
- 1.2 Evolution of Industrial Robots 14
  - Early Industrial Robots
- 1.3 What Is an Industrial Robot? 17
- 1.4 Types of Automation 18
  - Hard Automation • Flexible Automation
- 1.5 Are Robots a Threat? 21

Chapter 2: Fundamentals of Robotics 23
- 2.1 Parts of a Robot 24
  - Controller • Manipulator • End Effector • Power Supply • Means for Programming
- 2.2 Degrees of Freedom 30
- 2.3 Classifying Robots 34
  - Type of Control System • Type of Actuator Drive • Shape of Work Envelope

Chapter 3: Programming the Robot 59
- 3.1 The Evolution of Programming 60
  - First Generation • Second Generation • Third Generation
- 3.2 Motion Control 63
  - Pick-and-Place Motion • Point-to-Point Motion • Continuous-Path Motion
- 3.3 Programming Methods 67
  - Manual Programming • Using a Teach Pendant • Walk-Through Programming • Using a Computer Terminal
- 3.4 Programming Languages 70
- 3.5 Types of Programming 73
  - Hierarchical Control Programming • Task-Level Programming
- 3.6 Voice Recognition 77

Chapter 4: Industrial Applications 81
4.1 Integrating Robots into the Manufacturing Process 82
Design for Manufacturability • Robotic Safety Considerations
4.2 Selecting a Suitable Robot 89
Work Envelope • Degrees of Freedom • Resolution • Accuracy • Repeatability • Operations Speed • Load Capacity
4.3 Using Robots in Industry 93
Pick-and-Place • Machine Loading and Unloading • Die Casting • Welding • Spray Operations • Machine Processes • Assembly • Inspection • Material Handling • Service

Unit II—Power Supplies and Movement Systems

Chapter 5: Electromechanical Systems 115
5.1 Automated Systems and Subsystems 116
Energy Source • Transmission Path • Control • Load • Indicators
5.2 Mechanical Systems 119
5.3 Electrical Systems 119
Sensing Systems • Timing Systems • Control Systems • Rotary Motion Systems
5.4 Overload Protection 137

Chapter 6: Fluid Power Systems 139
6.1 Fluid Power System Models 140
Hydraulic System Model • Pneumatic System Model
6.2 Characteristics of Fluid Flow 144
Compression of Fluids
6.3 Principles of Fluid Power 146
Force, Pressure, Work, and Power
6.4 Fluid Power System Components 149
Fluid Pumps • Fluid Conditioning Devices • Transmission Lines • Control Devices • Load Devices • Indicators
6.5 Hybrid Systems 167

Unit III—Sensing and End-of-Arm Tooling

Chapter 7: Sensors 171
7.1 How Sensors Work 172
Transducers

7.2 Types of Sensors 175
    Light Sensors • Vision Sensors • Sound Sensors • Temperature • Sensors • Magnetic Field Sensors

7.3 Sensor Applications 189
    Speed Sensors • Mechanical Movement Sensors • Proximity Sensors

Chapter 8: End Effectors 195

8.1 End Effector Movement... 196
    Prehensile Movements • Nonprehensile Movements

8.2 Types of End Effectors.. 198
    Grippers • Tools

8.3 Changeable End Effectors... 204

8.4 End Effector Design... 205
    Desirable Characteristics • Custom-Designed End Effectors

Unit IV—Control Systems and Maintenance

Chapter 9: Computer Systems and Digital Electronics 211

9.1 Computer Systems.. 212
    Computer System Components • Basic Functions

9.2 Digital Number Systems 219
    Binary Number System • Octal Number System • Hexadecimal Number System

9.3 Binary Logic Circuits... 226
    Logic Gates • Flip-Flops • Digital Counters

9.4 Computer Programming... 236
    Instructions • Program Planning

Chapter 10: Interfacing and Vision Systems 241

10.1 Interfacing 242
    Digital Input Ports • Digital Output Ports • Serial Ports • Parallel Ports

10.2 Machine Vision 247
    Fundamentals of Machine Vision

Chapter 11: Maintaining Robotic Systems 253

11.1 Troubleshooting 254
    Troubleshooting Charts

11.2 General Servicing Techniques 256

11.3 Preventative Maintenance 261
  Developing a Maintenance Program • Implementing a New
  Program

Chapter 12: Robots in Modern Manufacturing 265

12.1 Using Robots in Manufacturing 266
  Productivity • Replacing Employees • The Pay Off • Successful
  Robotic Implementations

12.2 Evaluating Potential Uses for Robots... 269
  Matching the Technology to the Need • Considering the
  Advantages and Disadvantages • Non-Economic Justifications
  • Economic Justifications

12.3 Preparing an Implementation Plan... 275
  Identify Potential Applications • Analyze Potential Applications
  • Construct a Matrix for Comparing Equipment • Review the
  Available Equipment • Match the Robot to the Application
  • Develop the Proposal • Develop and Refine the Application
  • Begin Implementation • Provide Training • Provide Maintenance
  Programs

Chapter 13: The Future of Robotics 281

13.1 Fully-Automated Factories... 282
13.2 Robots Outside the Factory 282
13.3 Artificial Intelligence (AI) and Expert Systems 285
13.4 Impacts on Society 287
13.5 Your Future in Robotics 289

Glossary 291

Index 310

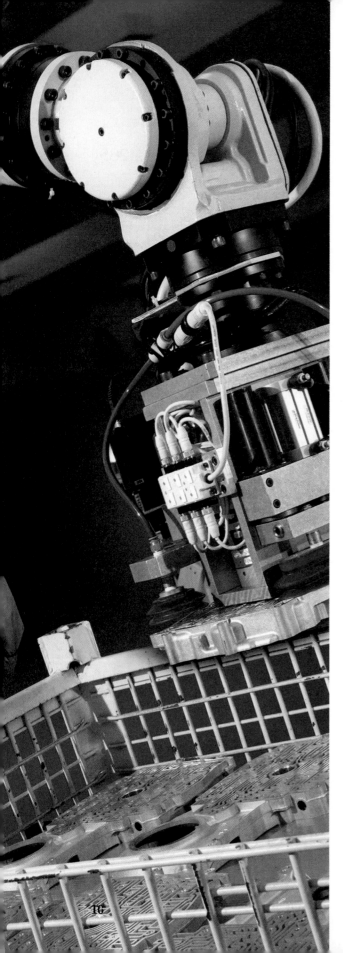

Unit I
Principles of Robotics

Since the first industrial robot was installed at a U.S. automotive plant in 1961, robotics technology has become an integral factor in most types of manufacturing. Robots are widely used for applications that require extreme precision, for repetitive and tedious tasks, and for work that is considered unpleasant or dangerous for humans. Robots are also vital components of flexible manufacturing systems, which allow robotic configurations to be quickly changed to meet production requirements.

Unit I Chapters

Chapter 1: Introduction to Industrial Robotics...11
1.1 Early Robots 12
1.2 Evolution of Industrial Robots 14
1.3 What Is an Industrial Robot? 17
1.4 Types of Automation 18
1.5 Are Robots a Threat? 21

Chapter 2: Fundamentals of Robotics...............23
2.1 Parts of a Robot 24
2.2 Degrees of Freedom 30
2.3 Classifying Robots 34

Chapter 3: Programming the Robot 59
3.1 The Evolution of Programming 60
3.2 Motion Control 63
3.3 Programming Methods 67
3.4 Programming Languages 70
3.5 Types of Programming 73
3.6 Voice Recognition 77

Chapter 4: Industrial Applications 81
4.1 Integrating Robots into the Manufacturing Process 82
4.2 Selecting a Suitable Robot 89
4.3 Using Robots in Industry 93

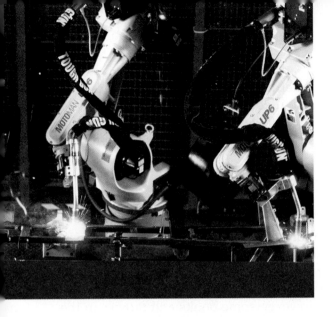

Chapter 1
Introduction to Industrial Robotics

Chapter Topics

1.1 Early Robots
1.2 First Industrial Robots
1.3 What Is an Industrial Robot?
1.4 Types of Automation
1.5 Are Robots a Threat?

Objectives

Upon completion of this chapter, you will be able to:
- Cite important developments in the evolution of robots.
- List and explain the classifications of industrial robots.
- Define the types of automation.
- Discuss the role of robots in our society.

Technical Terms

anthropomorphic	industrial robot	reprogrammable
artificial intelligence	intelligent robot	robot
automaton	manual manipulator	robotics
fixed-sequence robot	numerically controlled	Unimate
flexible automation	(NC) robot	variable-sequence
hard automation	playback robot	robot

Overview

For centuries, the idea of robots has captured people's imaginations. Who first conceived of robots? How did they become so popular? This chapter discusses the origins of the robot and explores the ways in which robots began to be used in industry.

11

The term *robot* is commonly defined as a machine or device that automatically performs tasks or activities which are typically completed using human skill and intelligence. However, this definition does not apply to every type of robot. This chapter also addresses the differences in how "robot" is defined.

1.1 Early Robots

Robots have long played important roles in the movies, in books, and on television. Characters like R2-D2 and C-3PO of the *Star Wars*© movies present robots in a favorable, comic light. Some real robots have a non-threatening appearance with familiar, even humanlike, characteristics and qualities (**Figure 1-1**). However, some books and movies in the past have portrayed robots as a threat to humankind, rather than a help. Exploring the origins of the robot will help in understanding why many people regarded robots as a threat.

The term *automaton* was originally used for what we now consider to be a robot. An automaton was a human-made object that moved automatically. The first useful automatons were clockworks introduced during the Middle Ages to automatically keep track of time. As these clocks became more complex, systems of gears and pulleys enabled the

Figure 1-1. These robots serve different purposes and vary in "humanlike" characteristics. A—This robot is used for educational activities and interacting with small children. (Courtesy of NEC Corporation. Unauthorized use not permitted.) B—This service robot may be used for video surveillance and security, and has the familiar characteristics of a family pet.

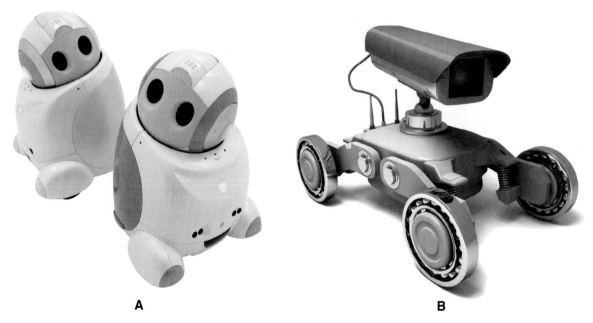

A   B

workings and figures attached to the clocks to move in lifelike ways. Many advanced automatons were built to entertain royalty and nobility because they appealed to people's imaginations. The English scientist Roger Bacon (ca. 1220–1292) even had ideas for a flying machine, a diving bell, a mechanical chariot, and mechanical birds.

As people became more knowledgeable about physiology, some believed that even the human body was merely a complex automaton separate from the mind and soul. The seventeenth-century French philosopher Ren Descartes (1596–1650) strongly supported this viewpoint, and people had visions of building truly lifelike automatons. However, since the machines would be mindless and soulless, would they wreak havoc on humans if they were to go out of control? The story is told that Descartes built a female automaton and took it on a sea voyage. The captain of the ship, thinking it was the work of the devil, promptly threw it overboard.

Robots in Literature

Mary Shelley's 1818 novel *Frankenstein* did little to ease people's fears of technology. Dr. Frankenstein's monster, an artificial man, was unhappy when his creator neglected him, so he took a horrible revenge.

In 1921, the Czech dramatist Karel Capek (1890–1938) wrote a play titled *R.U.R.* ("Rossum's Universal Robots"). In the play, Rossum develops a formula for making mechanical robots whose only function is to work. The robots are extremely strong and dedicated to their tasks; some are even quite intelligent, but they lack emotions. A young woman in the story objects to the inhumane treatment given the robots and persuades the physiologist at the plant to secretly change the formula for making the robots so that they become more human. As a result of this change, some robots develop interests other than work. The altered robots organize the original robots and turn them against humanity, eventually annihilating everyone except the company's founder.

The word "robot" is derived from the Czech word *robota*, which means forced labor. Capek's play was so popular that the word "robot" began to be used instead of "automaton" in virtually every language. Today, robots are commonly considered to be any manufactured structure that performs functions normally done by human beings.

Not all writers have portrayed robots in a negative manner. In 1939, when Isaac Asimov was only nineteen, he began to write science fiction in which robots were simply machines that could be built with safety measures in mind. Asimov's story *Runaround*, published in the March 1942 issue of *Astounding* magazine, established his three fundamental laws of robotics:

1. A robot may not injure a human being or, through inaction, allow a human being to come to harm.

2. A robot must obey the orders given it by human beings, except where such orders would conflict with the first law.

3. A robot must protect its own existence as long as such protection does not conflict with the first and second laws.

Much later, he added the Zeroth Law: A robot may not injure humanity or, through inaction, allow humanity to come to harm.

At the time, Asimov did not have a clear idea of how robots would actually be built. For instance, he described them as having "positronic," rather than electronic, brains. His original three laws of robotics, however, have been taken more seriously, and his term *robotics*, which means the use of robots, has become an established part of our language. Asimov's version of a robot is now the more generally accepted one, and Capek's version has faded into the past.

The Advent of Computers

By the end of World War II, computers began to be built. To many people, they seemed to be "thinking machines." Scientists combined the self-guided control systems developed during the war with computers. The idea of building humanlike machines came back into popularity. Both scientists and laypeople began to consider the consequences of merging computers with another new development, artificial intelligence. *Artificial intelligence* is the ability of a computer program to make decisions based on known information. Could a computer be installed into a structure resembling a human body and become a robotic human?

1.2 Evolution of Industrial Robots

The invention of the first *industrial robot*, a multifunction manipulator programmed to perform various tasks, was now close at hand. After the war, North American industry benefited from state-of-the-art equipment and technologies, while the other industrialized countries lay in shambles. By 1950, industry in the United States was at its peak. The demand for goods both at home and abroad was overwhelming, and the need for increased production through automated systems became apparent.

Early Industrial Robots

George C. Devol, Jr. patented the first industrial robot in 1954, but faced many problems in financing his invention and selling it to corporations in the United States. In 1956, Devol met a young engineer named Joseph Engelberger. Engelberger was impressed with Devol's idea and tried to persuade his employer, Aircraft Products, to help develop an industrial robot. However, it was not until Aircraft Products was acquired by Consolidated Diesel Electric Corporation that the necessary capital was made available.

The subsidiary Unimation, Inc. was formed in 1958, with Engelberger as president, to develop Devol's invention. In 1961, the first Unimation robot, called the *Unimate*, was sold to General Motors (**Figure 1-2**). Like many other early robots used in industry, the Unimate was not called a "robot" due to the many negative connotations. The first Unimate was used

Figure 1-2. The Unimate was the first industrial robot developed by Unimation in the 1950s.

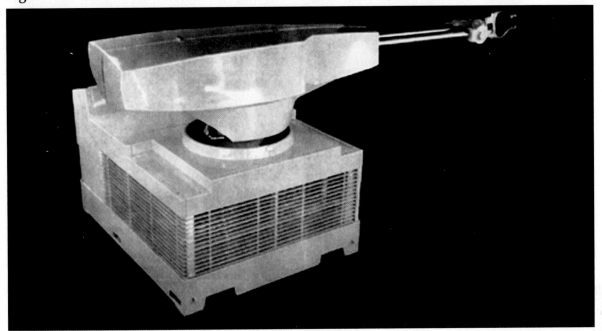

in a die-casting operation. The robot was first guided through the desired sequence of steps, which were recorded. The recording was then played back and the robot automatically performed the required task.

AMF Versatran built the manipulators used on atomic energy projects and became interested in robots around the same time Unimation was being formed. In 1959, AMF Versatran was purchased by Prab Conveyor Company, and together formed Prab Robots, Inc. Prab developed its own line of industrial robots in the late 1960s.

Other major manufacturers involved in early robot development for commercial use were DeVilbiss, Asea, and Cincinnati Milacron. DeVilbiss became one of the leading producers of finishing robots and introduced an arc-welding robot in 1982. Asea was one of the early developers of *anthropomorphic* (humanlike in form) robot units. Cincinnati Milacron also entered the robot market with an anthropomorphic unit powered by hydraulics.

During the late 1970s and early 1980s, robots moved into assembly operations. In 1978, engineers at Unimation introduced a smaller robot called PUMA (Programmable Universal Machine for Assembly). PUMAs were designed to handle small parts used in the assembly of motors and instruments.

Other American companies, such as IBM, Bendix, General Motors, and General Electric, eventually entered the robotics business. These companies offered foreign-built, but American-packaged, robots through various

Inventor Spotlight: George C. Devol

George C. Devol, Jr. has been called the "grandfather of industrial robots."

Prior to his collaboration with Joseph Engelberger to form the Unimation company in 1956, George Devol was part of many innovative developments.

- In 1932 George Devol formed the United Cinephone Corporation which manufactured phonograph arms and amplifiers, photoelectric doors called "Phantom Doorman," and developed a bar code system used in package sorting processes.
- The number of visitors at the 1939 World's Fair in New York was tracked using automated counters developed by United Cinephone Corporation.
- George Devol formed General Electronics Industries in 1943 and began developing counter-radar measures for military use. These systems were used during WWII by the allied forces.
- In 1946, George Devol patented a general-purpose playback device that used a magnetic process recorder. This early development led to programmable robotic systems.

licensing agreements with Japanese, German, and Italian companies. In 1982, a major joint venture was formed between General Motors and Fanuc of Japan to create GM Fanuc to market robots in North America.

Japan Enters the Market

The early success of robotics technology in the United States had not gone unnoticed by the Japanese. In 1966, many Japanese companies sent representatives to this country to see what was happening. In 1967, Joseph Engelberger was invited to tour Japan and he lectured in Tokyo to an audience of 700 engineers and executives. Robotics technology grew rapidly in that country.

Japan's first industrial robot was developed in 1969, after the first AMF Versatran robots had been exhibited and sold there. According to Engelberger, Japanese companies did not resist technology as American companies had, and Japan was able to enter the market quickly. The Japan Robot Association (JARA) was founded in 1971, three years before the Robotic Institute of America (now called the Robotic Industries Association or RIA), and six years before the British Automation and Robotics Association (BARA). In 1978, the SCARA (Selective Compliance Assembly Robot Arm) was developed at Yamanashi University and marketed by IBM and Sankyo. Japan became the world's largest user of robots.

1.3 What Is an Industrial Robot?

The most widely accepted definition for the term "industrial robot" in the United States has been published by the RIA: *A reprogrammable, multifunctional manipulator designed to move material, parts, tools, or specialized devices through various programmed motions for the performance of a variety of tasks.*

This definition contains several important points:

- The robot is a machine.
- The robot is *reprogrammable*, meaning that it can be given new instructions to meet changed requirements and perform new tasks.
- The robot has a multifunction manipulator, which means it may be used in different ways, even within the same program.
- The robot is flexible and can perform a variety of operations to meet special needs.

The robots pictured in **Figure 1-3** fulfill the requirements of the RIA definition. The robotic system consists of a motor-driven, multifunction

Figure 1-3. These industrial robots meet the RIA definition. (ABB Graco, Motoman, Fanuc Robotics)

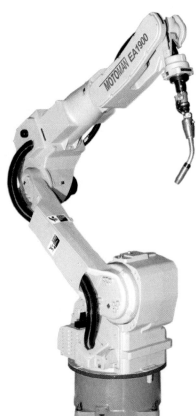
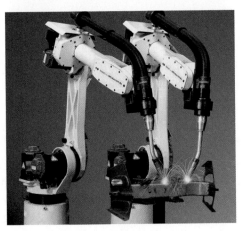

manipulator arm, an electronic memory system containing the program that controls manipulator movement, and a microcomputer for reprogramming the robot for new tasks. The progression of three generations of manufacturing robots is illustrated in **Figure 1-4**. It lists several applications used today and possible applications for the future. Many of these applications will be discussed in Chapter 4. Because of a robot's programmability, it can function in many different jobs. In the future, the majority of these jobs will be outside the manufacturing area.

As robots evolved, Japan categorized certain types of automated machinery, not considered robots in the United States, as industrial robots. This raised questions as to how many true robots were used by Japanese industry, compared to those used in the United States. The Japanese used a wide range of classifications, from simple arms to what they called "intelligent robots." These classifications include:

- *Manual manipulator.* A manipulator worked by a human operator.
- *Fixed-sequence robot.* A manipulator that repetitively performs successive steps of a given operation according to a predetermined sequence, condition, and position. Its instructions cannot be easily changed.
- *Variable-sequence robot.* A manipulator that is similar to the fixed-sequence robot, but its sequence of movement can be easily changed.
- *Playback robot.* A manipulator that can reproduce operations originally executed under human control. An operator initially feeds in the instructions relating to sequence of movement, conditions, and positions. These are then stored in memory.
- *Numerically controlled (NC) robot.* A manipulator that can perform the sequence of movement, conditions, and positions of a given task which are communicated by means of numerical data.
- *Intelligent robot.* A robot that can, by itself, detect changes in the work environment by means of sensors (visual and/or tactile). Using its decision-making capability, it can then proceed with the appropriate operations.

1.4 Types of Automation

Today's industries use various types of automation to manufacture parts and products. Two common classifications are hard automation and flexible automation.

Hard Automation

Hard automation refers to machinery that has been specifically designed and built to perform particular tasks within an assembly line. This approach works well when a very large number of items are to be produced over a long period of time. After an item the machine was designed to

Figure 1-4. Three generations of industrial robots show increasing ability to accomplish more difficult tasks.

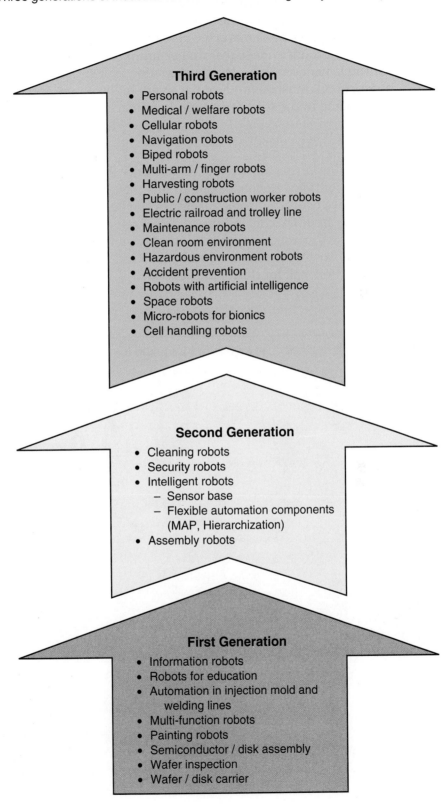

make is no longer needed, the machine is often discarded due to the high cost of retooling. Even though this kind of automation can be very costly, it is very reliable and the cycle times and machine precision is generally high. With the increasing demand for new products and new models of existing products, product life spans are becoming shorter and shorter. To justify the implementation of a hard automation system in today's industry, the product built must be produced at a very high volume and have the potential of being in use for a long period of time.

Flexible Automation

Flexible automation includes machines that are capable of performing a variety of tasks. Robots belong in this category. As new products or new models are needed, flexible machines can be reprogrammed to make the parts required. This flexibility can save money because equipment does not have to be discarded or rebuilt. In addition, it takes much less time to reprogram an existing machine than to build and install a new one. **Figure 1-5** shows a work cell for flexible automation.

Figure 1-5. A flexible manufacturing work cell can be programmed for more than one task.

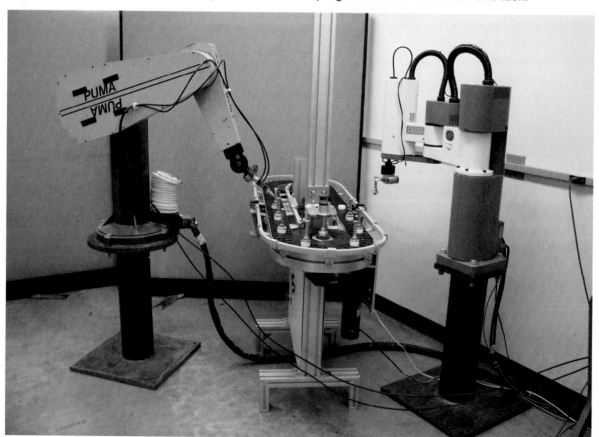

Robots are the most flexible type of automation available today. Not only can they be reprogrammed easily, quickly, and economically, but they can also be moved from one location in a plant to another.

1.5 Are Robots a Threat?

Our thinking has come a long way since Karel Capek wrote his play in 1920. However, some of his ideas are still being considered. Robots are more productive than humans. They can perform work less expensively and have taken jobs away from humans. For these reasons, workers and labor unions often see robots as a threat. At the same time, robots are performing work that relieves men and women from repetitive operations that are not only monotonous, but may also be unpleasant and dangerous, **Figure 1-6**. The addition of robots in the workforce does not necessarily mean that jobs have been lost. Technicians and other types of skilled laborers are needed to install, program, and maintain the robots.

Perhaps it is best to keep the words of Joseph Engelberger in mind: "Nobody needs a robot. There isn't anything that a robot can do that a willing human being can't do better."

Figure 1-6. The SUGV 300 (small unmanned ground vehicle) is a remote-controlled robot used in military operations. This robot can be sent into hazardous or inhospitable areas for inspection or tactical maneuvers. (iRobot Corporation)

Review Questions

Write your answers on a separate sheet of paper. Do not write in this book.

1. Who was the dramatist that introduced the word "robot" in his writing? What is the Czech meaning of the word *robota*?
2. Identify Isaac Asimov's first three laws of robotics and discuss their significance.
3. What year was the first industrial robot patented?
4. Who were George Devol and Joseph Engelberger?
5. Some robots are considered *anthropomorphic*. What does this term mean?
6. Robots are considered to be a key element in automated manufacturing. Define the term *industrial robot* and discuss the key factors in that definition.
7. List and define the six Japanese classifications of robots. Which of the classifications best fits the Robotic Industries Association's definition of an industrial robot?
8. List advantages and disadvantages for both hard automation and flexible automation.
9. Some people believe that the use of automation and robots will replace human workers. Others believe that if automation and robots are not incorporated into manufacturing, we will lose a far greater number of jobs. Explain how both views can be defended.

Learning Extensions

1. Visit Robotics Online (www.robotics.org) and the Society of Manufacturing Engineers Robotics International (www.sme.org). Research the impact of robotics in our current society and investigate future developments and innovations.
2. Locate the Web sites for at least three robotics-related professional and student associations.
3. Create a timeline that charts important dates, people, developments, and writings in the evolution of robots.

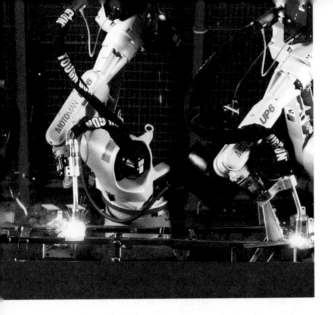

Chapter 2
Fundamentals of Robotics

Chapter Topics

2.1 Parts of a Robot
2.2 Degrees of Freedom
2.3 Classifying Robots

Objectives

Upon completion of this chapter, you will be able to:
- Identify the parts of a robot.
- Explain degrees of freedom.
- Discuss the difference between servo and non-servo robots.
- Identify and explain the different robot configurations.

Technical Terms

actuator
Cartesian
   configuration
closed-loop system
controller
cylindrical
   configuration
degrees of freedom
direct-drive motor
end effector
error signal
hierarchical control

hydraulic drive
linear actuator
manipulator
non-servo robot
open-loop system
pitch
pneumatic drive
power supply
program
radial traverse
revolute configuration
roll

rotary actuator
rotational traverse
SCARA
servo amplifier
servo robot
spherical configuration
tachometer
teach pendant
trajectory
vertical traverse
work envelope
yaw

23

Overview

Even the most complex robotic system can be broken down into a few basic components, which provide an overview of how a robot works. These components are covered in this chapter, with more detail provided in later chapters. Freedom of motion and the resulting shape of the robot's work area are also addressed in this chapter.

2.1 Parts of a Robot

Robots come in many shapes and sizes. The industrial robots illustrated in **Figure 2-1** resemble an inverted human arm mounted on a base. Robots consist of a number of components, **Figure 2-2**, that work together: the controller, the manipulator, an end effector, a power supply, and a means for programming. The relationship among these five components is illustrated in **Figure 2-3**.

Figure 2-1. This robot has been designed expressly for use in precise path-oriented tasks such as deburring, milling, sanding, gluing, bonding, cutting, and assembly. (Reis Machines, Inc.)

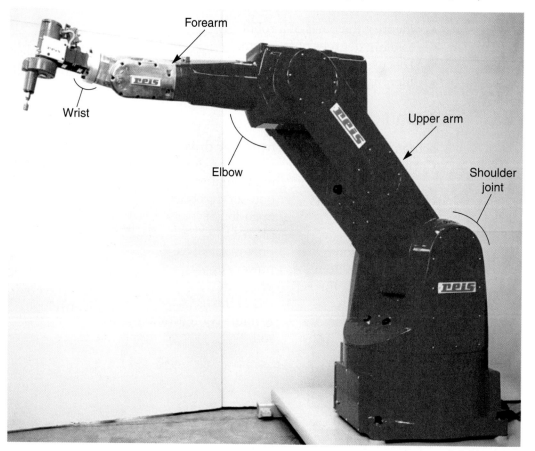

Figure 2-2. This robot illustrates the systems of a typical industrial robot. This electric robot can be used in a variety of industrial applications. (ABB Robotics)

Figure 2-3. The relationships among the five major systems that make up an industrial robot are shown in this diagram.

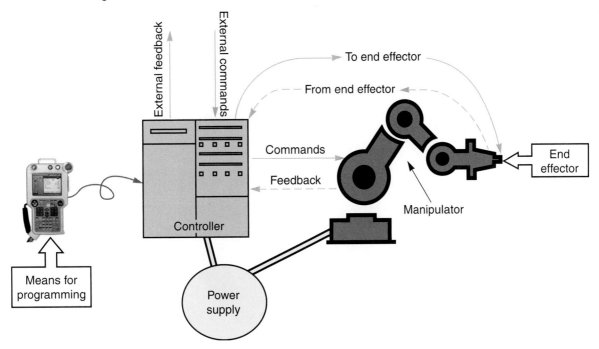

Controller

The *controller* is the part of a robot that coordinates all movements of the mechanical system, **Figure 2-4**. It also receives input from the immediate environment through various sensors. The heart of the robot's controller is generally a microprocessor linked to input/output and monitoring devices. The commands issued by the controller activate the motion control mechanism, consisting of various controllers, amplifiers, and actuators. An *actuator* is a motor or valve that converts power into robot movement. This movement is initiated by a series of instructions, called a *program*, stored in the controller's memory.

The controller has three levels of *hierarchical control*. Hierarchical control assigns levels of organization to the controllers within a robotic system. Each level sends control signals to the level below and feedback signals to the level above. The levels become more elemental as they progress toward the actuator. Each level is dependent on the level above it for instructions, **Figure 2-5**.

Figure 2-4. A controller/power supply with a teach pendant. (Motoman)

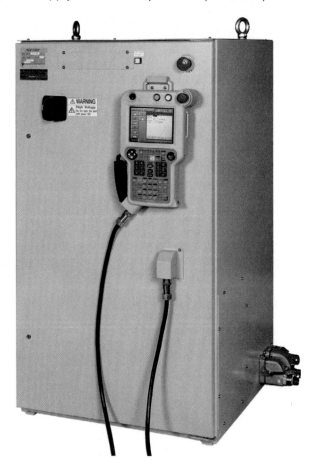

Figure 2-5. The three basic levels of hierarchical control.

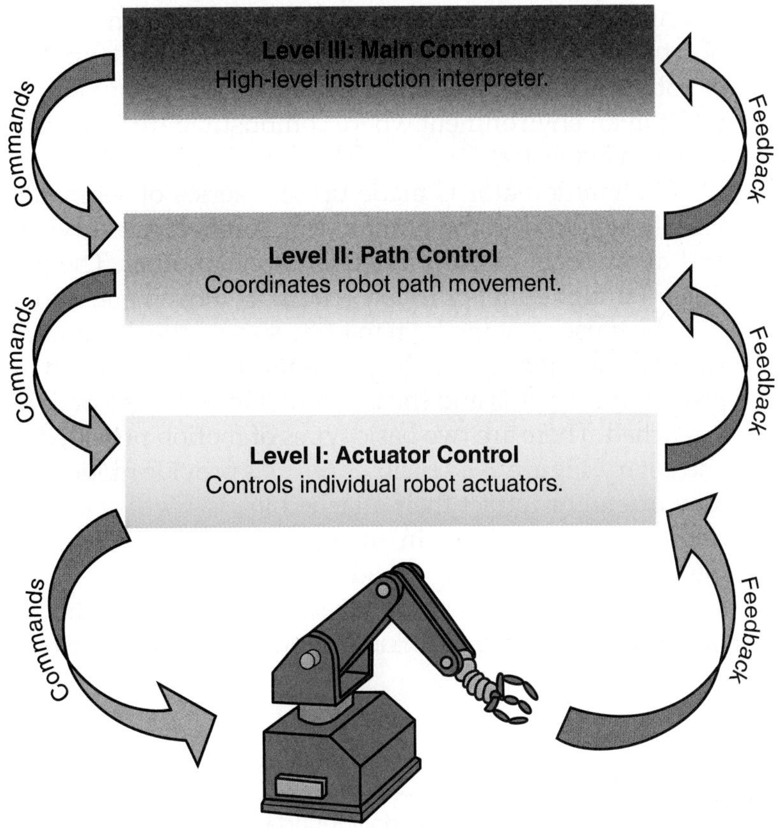

The three levels are:
- **Level I—Actuator Control.** The most elementary level at which separate movements of the robot along various planes, such as the X, Y, and Z axes, are controlled. These movements will be explained in detail later in this chapter.
- **Level II—Path Control.** The path control (intermediate) level coordinates the separate movements along the planes determined in Level I into the desired *trajectory* or path.
- **Level III—Main Control.** The primary function of this highest control level is to interpret the written instructions from the human programmer regarding the tasks required. The instructions are then combined with various environmental signals and translated by the controller into the more elementary instructions that Level II can understand.

Manipulator

The manipulator consists of segments that may be jointed and that move about, allowing the robot to do work. The *manipulator* is the arm of the robot (see **Figures 2-2** and **2-3**) which must move materials, parts, tools, or special devices through various motions to provide useful work. A manipulator can be identified by method of control, power source, actuation of the joints, and other factors. These factors help identify the best type of robot for the task at hand. For example, you would not use an electric robot in an environment where combustible fumes exist and a spark could cause an explosion.

The manipulator is made up of a series of segments and joints much like those found in the human arm. Joints connect two segments together and allow them to move relative to one another. The joints provide either linear (straight line) or rotary (circular) movement, **Figure 2-6**.

The muscles of the human body supply the driving force that moves the various body joints. Similarly, a robot uses actuators to move its arm along programmed paths and then to hold its joints rigid once the correct position is reached. There are two basic types of motion provided by actuators: linear and rotary, **Figure 2-7**. *Linear actuators* provide motion along a straight line; they extend or retract their attached loads. *Rotary actuators* provide rotation, moving their loads in an arc or circle. Rotary motion can be converted into linear motion using a lead screw or other mechanical means of conversion. These types of actuators are also used outside the robot to move workpieces and provide other kinds of motion within the work envelope.

Figure 2-6. Both linear and rotary joints are commonly found in robots.

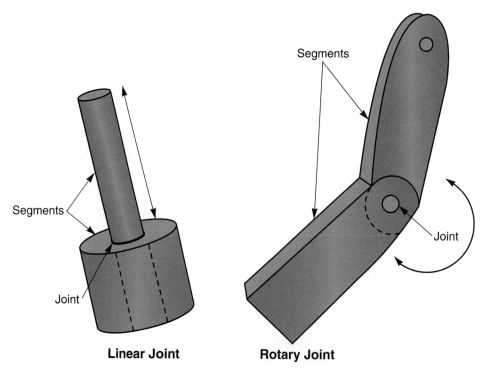

Figure 2-7. Actuators can be powered by electric motors, pneumatic (air) cylinders, or hydraulic (oil) cylinders. Linear actuators provide straight-line movement. Rotational movement around an axis is provided by the angular (rotary) actuator. (PHD, Inc.)

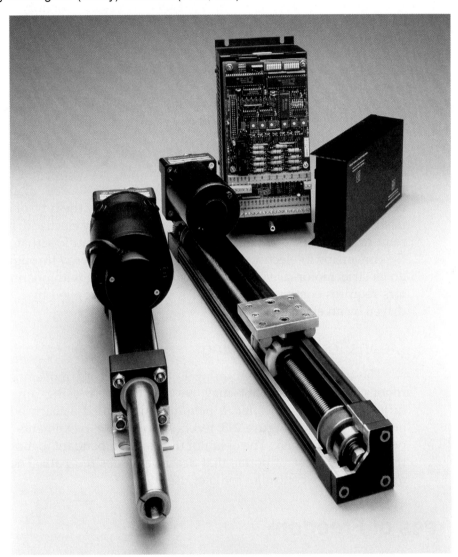

A *tachometer* is a device used to measure the speed of an object. In the case of robotic systems, a tachometer is used to monitor acceleration and deceleration of the manipulator's movements.

End Effector

The *end effector* is the robot's hand, or the end-of-arm tooling on the robot. It is a device attached to the wrist of the manipulator for the purpose of grasping, lifting, transporting, maneuvering, or performing operations on a workpiece. The end effector is one of the most important components of a robot system. The robot's performance is a direct result of how well the end effector meets the task requirements. The area within reach of the robot's end effector is called its *work envelope*.

Power Supply

The *power supply* provides the energy to drive the controller and actuators. It may convert ac voltage to the dc voltage required by the robot's internal circuits, or it may be a pump or compressor providing hydraulic or pneumatic power. The three basic types of power supplies are electrical, hydraulic, and pneumatic.

The most common energy source available, where industrial robots are used, is electricity. The second most common is compressed air, and the least common is hydraulic power. These primary sources of energy must be converted into the form and amount required by the type of robot being used. The electronic part of the control unit, and any electric drive actuator, requires electrical power. A robot containing hydraulic actuators requires the conversion of electrical power into hydraulic energy through the use of an electric, motor-driven, hydraulic pump. A robot with pneumatic actuators requires compressed air, which is usually supplied by a compressor driven by an electric motor.

Means for Programming

The means for programming is used to record movements into the robot's memory. A robot may be programmed using any of several different methods. The *teach pendant*, also called a teach box or hand-held programmer, **Figure 2-8**, teaches a robot the movements required to perform a useful task. The operator uses a teach pendant to move the robot through the series of points that describe its desired path. The points are recorded by the controller for later use.

2.2 Degrees of Freedom

Although robots have a certain amount of dexterity, it does not compare to human dexterity. The movements of the human hand are controlled by 35 muscles. Fifteen of these muscles are located in the forearm. The arrangement of muscles in the hand provides great strength to the fingers and thumb for grasping objects. Each finger can act alone or together with the thumb. This enables the hand to do many intricate and delicate tasks. In addition, the human hand has 27 bones. **Figure 2-9** shows the bones found in the hand and wrist. This bone, joint, and muscle arrangement gives the hand its dexterity.

Degrees of freedom (DOF) is a term used to describe a robot's freedom of motion in three dimensional space—specifically, the ability to move forward and backward, up and down, and to the left and to the right. For each degree of freedom, a joint is required. A robot requires six degrees of

Figure 2-8. This teach pendant is connected to a controller and is used to teach a robot how to complete a task. (Motoman)

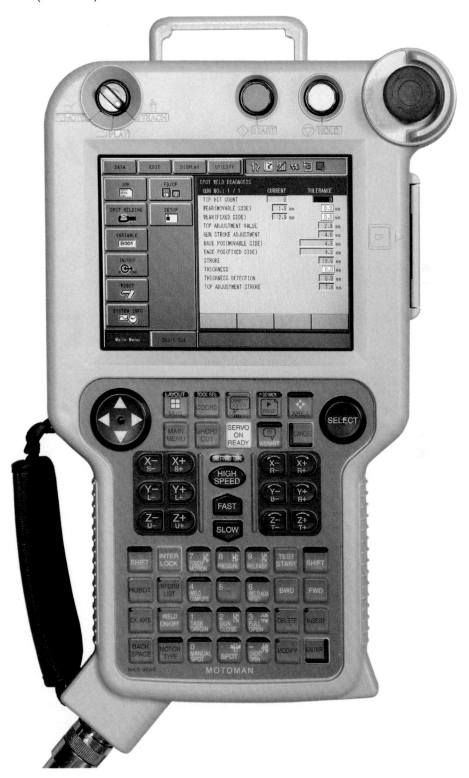

Figure 2-9. The arrangement of bones and joints found in the human hand provides dexterity. Each joint represents a degree of freedom; there are 22 joints, and thus, 22 degrees of freedom in the human hand.

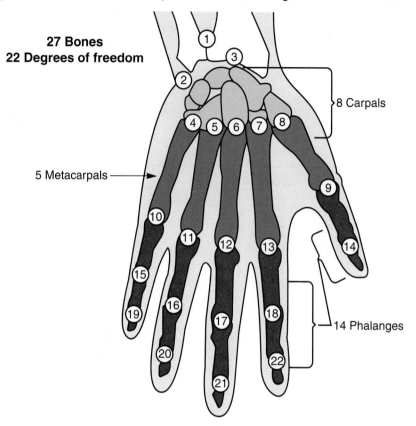

freedom to be completely versatile. Its movements are clumsier than those of a human hand, which has 22 degrees of freedom.

The number of degrees of freedom defines the robot's configuration. For example, many simple applications require movement along three axes: X, Y, and Z. See **Figure 2-10**. These tasks require three joints, or three degrees of freedom. The three degrees of freedom in the robot arm are the rotational traverse, the radial traverse, and the vertical traverse. The *rotational traverse* is movement on a vertical axis. This is the side-to-side swivel of the robot's arm on its base. The *radial traverse* is the extension and retraction of the arm, creating in-and-out motion relative to the base. The *vertical traverse* provides up-and-down motion.

For applications that require more freedom, additional degrees can be obtained from the wrist, which gives the end effector its flexibility. The three degrees of freedom in the wrist have aeronautical names: pitch, yaw, and roll. See **Figure 2-11**. The *pitch*, or bend, is the up-and-down movement of the wrist. The *yaw* is the side-to-side movement, and the *roll*, or swivel, involves rotation.

Figure 2-10. The three basic degrees of freedom are associated with movement along the X, Y, and Z axes of the Cartesian coordinate system.

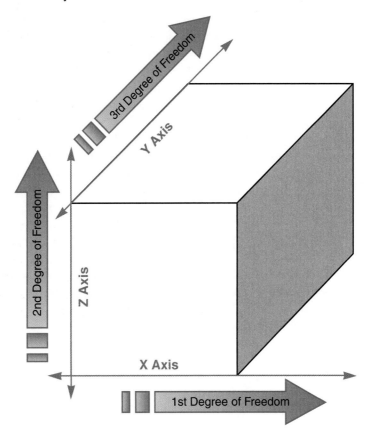

Figure 2-11. Three additional degrees of freedom—pitch, yaw, and roll—are associated with the robot's wrist. (Mack Corporation)

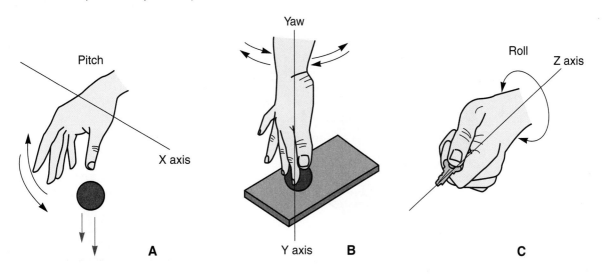

A robot requires a total of six degrees of freedom to locate and orient its hand at any point in its work envelope, **Figure 2-12**. Although six degrees of freedom are required for maximum flexibility, most applications require only three to five. When more degrees of freedom are required, the robot's motions and controller design become more complex. Some industrial robots have seven or eight degrees of freedom. These additional degrees are achieved by mounting the robot on a track or moving base, as shown in **Figure 2-13**. The track-mounted robot shown in **Figure 2-14** has a total of seven degrees of freedom. This addition also increases the robot's reach.

Although the robot's freedom of motion is limited in comparison with that of a human, the range of movement in each of its joints is considerably greater. For example, the human hand has a bending range of only about 165 degrees. The illustrations in **Figure 2-15** show the six major degrees of freedom by comparing those of a robot to a person using a spray gun.

2.3 Classifying Robots

Robots can be classified in various ways, depending on their components, configuration, and use. Three common methods of classifying robots are by the types of control system used, the type of actuator drive used, and the shape of the work envelope.

Figure 2-12. Six degrees of freedom provide maximum flexibility for an industrial robot.

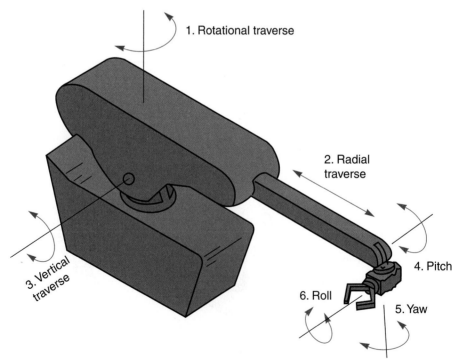

Figure 2-13. Using a gantry robot creates a large work envelope (A) because the manipulator arm is mounted on tracks (B). (Schunk)

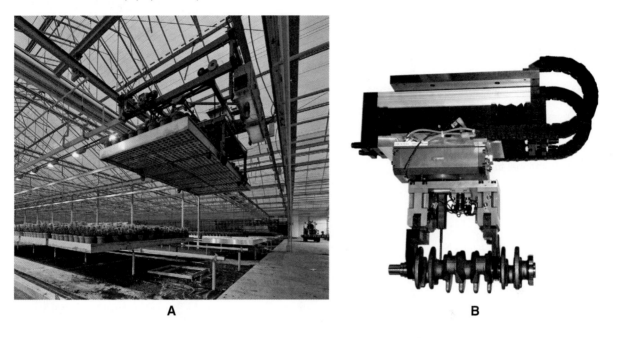

A B

Figure 2-14. Mounting this robot on tracks gives the system seven degrees of freedom—six from the configuration of the robot and one additional degree from the track mount.

Figure 2-15. The six degrees of freedom, demonstrated by a person using a spray gun. Illustrations 1, 2, and 3 are arm movements. Illustrations 4, 5, and 6 are wrist movements.

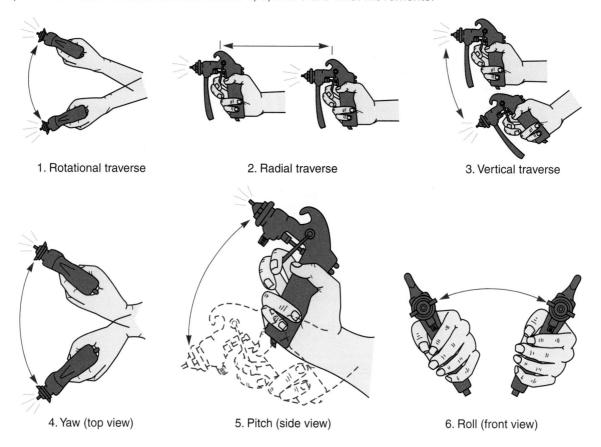

1. Rotational traverse
2. Radial traverse
3. Vertical traverse
4. Yaw (top view)
5. Pitch (side view)
6. Roll (front view)

Type of Control System

Robots may use one of two control systems—non-servo and servo. The earliest type of robot was non-servo, which is considered a non-intelligent robot. The second classification is the servo robot. These robots are classified as either intelligent or highly intelligent. The primary difference between an intelligent and highly intelligent robot is the level of awareness of its environment.

Non-Servo Robots

Non-servo robots are the simplest robots and are often referred to as "limited sequence," "pick-and-place," or "fixed-stop robots." The non-servo robot is an open-loop system. In an *open-loop system*, no feedback mechanism is used to compare programmed positions to actual positions.

A good example of an open-loop system is the operating cycle of a washing machine, **Figure 2-16**. At the beginning of the operation, the dirty clothes and the detergent are placed in the machine's tub. The cycle selector is set for the proper

Figure 2-16. This block diagram depicts the sequence of steps performed by a washing machine. Notice that no feedback is used. In such an open-loop control system, the condition of the clothes during the washing operation is not monitored and used to alter the process.

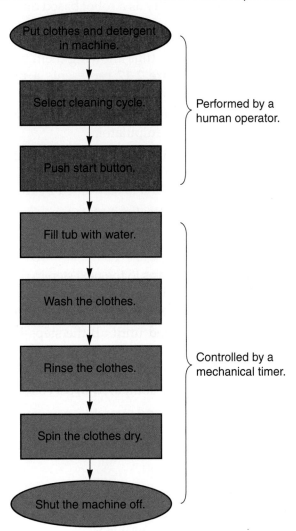

cleaning cycle and the machine is activated by the start button. The machine fills with water and begins to go through the various washing, rinsing, and spinning cycles. The machine finally stops after the set sequence is completed. The washing machine is considered an open-loop system for two reasons:

- The clothes are never examined by sensors during the washing cycle to see if they are clean.
- The length of the cycle is not automatically adjusted to compensate for the amount of dirt remaining in the clothes. The cycle and its time span are determined by the fixed sequence of the cycle selector.

Non-servo robots are also limited in their movement and these limitations are usually in the form of a mechanical stop. This form of robot is excellent in repetitive tasks, such as material transfer. One may question if the non-servo robots qualify as a robot based on the definition provided by the Robot Institute of America. However, if these robots are equipped with a programmable logic controller (PLC) they easily meet the requirement of a reprogrammable device, thus allowing them to be classified as a robot.

The diagram in **Figure 2-17** represents a pneumatic (air-controlled), non-servo robot.

1. At the beginning of the cycle, the controller sends a signal to the control valve of the manipulator.
2. As the valve opens, air passes into the air cylinder, causing the rod in the cylinder to move. As long as the valve remains open, this rod continues to move until it is restrained by the end stop.
3. After the rod reaches the limit of its travel, a limit switch tells the controller to close the control valve.
4. The controller sends the control valve a signal to close.
5. The controller then moves to the next step in the program and initiates the necessary signals. If the signals go to the robot's end effector, for example, they might cause the gripper to close in order to grasp an object.

The process is repeated until all the steps in the program have been completed.

Characteristics of non-servo robots:

- Relatively inexpensive compared to servo robots.
- Simple to understand and operate.
- Precise and reliable.

Figure 2-17. In a non-servo system, movement is regulated by devices such as a limit switch, which signals the controller when it is activated.

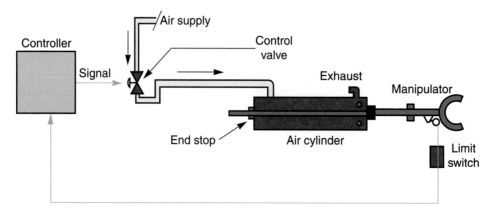

- Simple to maintain.
- Capable of fairly high speeds of operation.
- Small in size.
- Limited to relatively simple programs.

Servo Robots

The *servo robot* is a closed-loop system because it allows for feedback. In a *closed-loop system*, the feedback signal sent to the *servo amplifier* affects the output of the system. A servo amplifier translates signals from the controller into motor voltage and current signals. Servo amplifiers are used in motion control systems where precise control of position or velocity is necessary. In a sense, a servomechanism is a type of control system that detects and corrects for errors. **Figure 2-18** shows a block diagram of a servo robot system.

The principle of servo control can be compared to many tasks performed by human beings. One example is cutting a circle from a piece of stock on a power bandsaw, shown in **Figure 2-19**. The machine operator's eye studies the position of the stock to be cut in relation to the cutting edge of the blade. The eye transmits a signal to the brain. The brain compares the actual position to the desired position. The brain then sends a signal to the arms to move the stock beneath the cutting edge of the blade. The eyes are used as a feedback sensing device, while the brain compares desired locations with

Figure 2-18. A servo robot system, such as the one depicted in this block diagram, might be classified as intelligent or highly intelligent, depending on the level of sensory data it can interpret. Components in the shaded area are part of the control system.

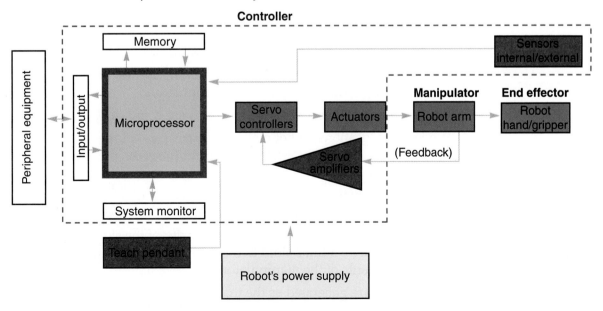

Figure 2-19. Human beings make use of the servomechanism principle for many tasks, such as cutting a circle on a bandsaw.

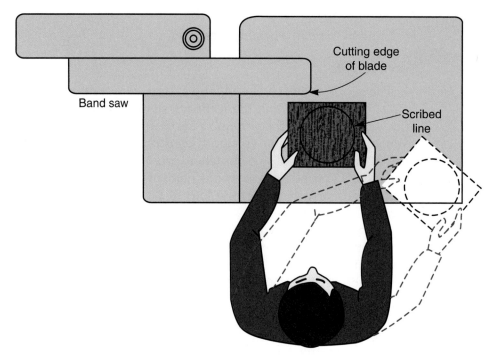

actual locations. The brain sends signals to the arms to make necessary adjustments. This process is repeated as the operator follows the scribed line during the sawing operation.

The diagram in **Figure 2-20** details one of the axes used in a hydraulic robot and helps to explain its operation.

1. When the cycle begins, the controller searches the robot's programming for the desired locations along each axis.
2. Using the feedback signals, the controller determines the actual locations on the various axes of the manipulator.
3. The desired locations and actual locations are compared.
4. When these locations do not match, an *error signal* is generated and fed back to the servo amplifier. The greater the error, the higher the intensity of the signal.
5. These error signals are increased by the servo amplifier and applied to the control valve on the appropriate axis.
6. The valve opens in proportion to the intensity of the signal received. The opened valve admits fluid to the proper actuator to move the various segments of the manipulator.
7. New signals are generated as the manipulator moves.

Figure 2-20. Feedback signals from sensors allow the system to make corrections whenever the actual speed or position of the robot does not agree with the values contained in the robot's program.

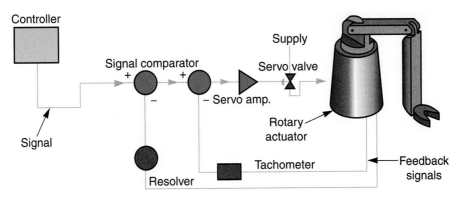

8. The servo control valves close when there are no more error signals, shutting off the flow of fluid.
9. The manipulator comes to rest at the desired position.
10. The controller then addresses the next instruction in the program, which may be to move to another location or operate some peripheral equipment.

The process is repeated until all steps of the program are completed. Characteristics of servo robots:

- Relatively expensive to purchase, operate, and maintain.
- Use a sophisticated, closed-loop controller.
- Wide range of capabilities.
- Can transfer objects from one point to another, as well as along a controlled, continuous path.
- Respond to very sophisticated programming.
- Use a manipulator arm that can be programmed to avoid obstructions within the work envelope.

Type of Actuator Drive

One common method of classifying robots is the type of drive required by the actuators.

- Electrical actuators use electric power.
- Pneumatic actuators use pneumatic (air) power.
- Hydraulic actuators, **Figure 2-21**, use hydraulic (fluid) power.

Figure 2-21. A large hydraulic actuator provides up-and-down motion to the manipulator arm of this industrial robot. (FANUC Robotics)

Electric Drive

Three types of motors are commonly used for electric actuator drives: ac servo motors, dc servo motors, and stepper motors. Both ac and dc servo motors have built-in methods for controlling exact position. Many newer robots use servo motors rather than hydraulic or pneumatic ones. Small and medium-size robots commonly use dc servo motors. Because of their high torque capabilities, ac servo motors are found in heavy-duty robots, **Figure 2-22**. A stepper motor is an incrementally controlled dc motor. Stepper motors are rarely used in commercial industrial robots, but are commonly found in educational robots, **Figure 2-23**.

Conventional, electric-drive motors are quiet, simple, and can be used in clean-air environments. Robots that use electric actuator drives require less floor space, and their energy source is readily available. However, the conventionally geared drive causes problems of backlash, friction, compliance, and wear. These problems cause inaccuracy, poor dynamic response, need for regular maintenance, poor torque control capability, and limited maximum speed on longer moves. Loads that are heavy enough to stall (stop) the motor can cause damage. Conventional electric-drive motors also have poor output power compared to their weight. This means that a larger, heavier motor must be mounted on the robot arm when a large amount of torque is needed.

Figure 2-22. This heavy-duty industrial robot uses two ac servo motors in the operation of its manipulator arm. (Motoman)

— AC servo motor

— AC servo motor

Figure 2-23. DC stepper motors are used on this tabletop educational robot. (Techno, Inc.)

The rotary motion of most electric actuator drives must be geared down (reduced) to provide the speed or torque required by the manipulator. However, manufacturers are beginning to offer robots that use *direct-drive motors*, which eliminate some of these problems. These high-torque motors drive the arm directly, without the need for reducer gears. The prototype of a direct-drive arm was developed by scientists at Carnegie-Mellon University in 1981.

The basic construction of a direct-drive motor is shown in **Figure 2-24**. Coupling the motor with the arm segment to be manipulated eliminates backlash, reduces friction, and increases the mechanical stiffness of the drive mechanism. Compare the design of a robot arm using a direct-drive motor in **Figure 2-25** to one with a conventional electric-drive (**Figure 2-22**). Using direct-drive motors in robots results in a more streamlined design. Maintenance requirements are also reduced. Robots that use direct-drive motors operate at higher speeds, with greater flexibility, and greater accuracy than those that use conventional electric-drive motors.

Applications currently being performed by robots with direct-drive motors are mechanical assembly, electronic assembly, and material handling. These robots will increasingly meet the demands of advanced, high-speed, precision applications.

Hydraulic Drive

Many earlier robots were driven by hydraulic actuator drives. A *hydraulic drive* system uses fluid and consists of a pump connected to a reservoir tank,

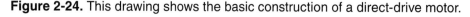

Figure 2-24. This drawing shows the basic construction of a direct-drive motor.

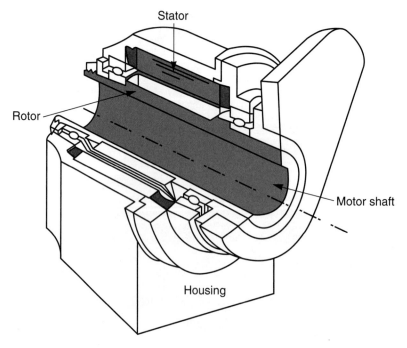

Figure 2-25. Note the simplified mechanical design of this direct-drive robot as compared to the conventional electric-drive robot shown in Figure 2-22. (Adept Technology, Inc.)

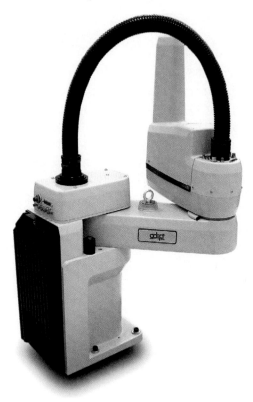

control valves, and a hydraulic actuator. Hydraulic drive systems provide both linear and rotary motion using a much simpler arrangement than conventional electric-drive systems, **Figure 2-26**. The storage tank supplies a large amount of instant power, which is not available from electric-drive systems.

Hydraulic actuator drives have several advantages. They provide precise motion control over a wide range of speeds. They can handle heavy loads on the end of the manipulator arm, can be used around highly explosive materials, and are not easily damaged when quickly stopped while carrying a heavy load. However, they are expensive to purchase and maintain and are not energy efficient. Hydraulic actuator drivers are also noisier than electric-drive actuators and are not recommended for clean-room environments due to the possibility of hydraulic fluid leaks.

Pneumatic Drive

Pneumatic drive systems make use of air-driven actuators. Since air is also a fluid, many of the same principles that apply to hydraulic systems are applicable to pneumatic systems. Pneumatic and hydraulic motors and cylinders are very similar. Since most industrial plants have a compressed

Figure 2-26. Large robots that use hydraulic drive systems perform a demonstration at a manufacturing trade show. (ABB Robotics)

air system running throughout assembly areas, air is an economical and readily available energy source. This makes the installation of robots that use pneumatic actuator drives easier and less costly than that of hydraulic robots. For lightweight pick-and-place applications that require both speed and accuracy, a pneumatic robot is potentially a good choice.

Pneumatic actuator drives work at high speeds and are most useful for small-to-medium loads. They are economical to operate and maintain and can be used in explosive atmospheres. However, since air is compressible, precise placement and positioning require additional components to achieve the smooth control possible with a hydraulic system. These components are discussed in later chapters. It is also difficult to keep the air as clean and dry as the control system requires. Robots that use pneumatic actuator drives are noisy and vibrate as the air cylinders and motors stop.

Shape of the Work Envelope

Robots come in many sizes and shapes. The type of coordinate system used by the manipulator also varies. The type of coordinate system, the arrangement of joints, and the length of the manipulator's segments all help determine the shape of the work envelope. To identify the maximum work area, a point on the robot's wrist is used, rather than the tip of the gripper or the end of the tool bit. Therefore, the work envelope is slightly larger when the tip of the tool is considered.

Work envelopes vary from one manufacturer to another, depending on the exact design of the manipulator arm. Combining different configurations in a single robot can result in another set of possible work envelopes. Before choosing a particular robot configuration, the application must be studied carefully to determine the precise work envelope requirements.

Some work envelopes have a geometric shape; others are irregular. One method of classifying a robot is by the configuration of its work envelope. Some robots may be equipped for more than one configuration. The four major configurations are: revolute, Cartesian, cylindrical, and spherical. Each configuration is used for specific applications.

Revolute Configuration (Articulated)

The *revolute configuration*, or jointed-arm, is the most common. These robots are often referred to as anthropomorphic because their movements closely resemble those of the human body. Rigid segments resemble the human forearm and upper arm. Various joints mimic the action of the wrist, elbow, and shoulder. A joint called the *sweep* represents the waist.

A revolute coordinate robot performs in an irregularly shaped work envelope. There are two basic revolute configurations: vertically articulated and horizontally articulated.

The vertically articulated configuration, shown in **Figure 2-27**, has five revolute (rotary) joints. A vertically articulated robot is depicted in **Figure 2-28**. The jointed-arm, vertically articulated robot is useful for painting applications because of the long reach this configuration allows.

The horizontally articulated configuration generally has one vertical (linear) and two revolute joints. Also called the *SCARA* (selective compliance assembly robot arm) configuration, it was designed by Professor Makino of Yamanashi University, Japan. The primary objective was a configuration that would be fairly yielding in horizontal motions and rather rigid in vertical motions. The basic SCARA configuration, **Figure 2-29**, is an adaptation of the cylindrical configuration. The SCARA robot shown in **Figure 2-30** is designed for clean-room applications, such as wafer and disk handling in the electronics industry.

SCARA robots are ideally suited for operations in which the vertical motion requirements are small compared to the horizontal motion requirements. Such an application would be assembly work where parts are picked up from a parts holder and moved along a nearly horizontal path to the unit being assembled.

The revolute configuration has several advantages. It is, by far, the most versatile configuration and provides a larger work envelope than the Cartesian, cylindrical, or spherical configurations. It also offers a more flexible reach than the other configurations, making it ideally suited to welding and spray painting operations.

However, there are also disadvantages to the revolute configuration. It requires a very sophisticated controller, and programming is more complex than for the other three configurations. Different locations in the work envelope

Figure 2-27. These five revolute (rotary) joints are associated with the basic manipulator movements of a vertically articulated robot. (Adept Technology, Inc.)

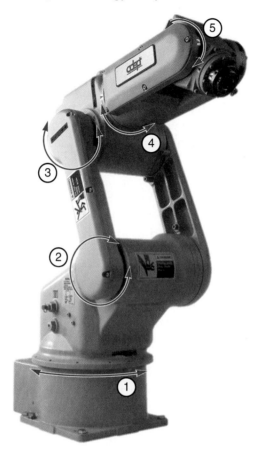

Figure 2-28. A—This painting robot is vertically articulated. (ABB Graco Robotics, Inc.) B—The shaded areas represent a top view of the work envelope for this robot.

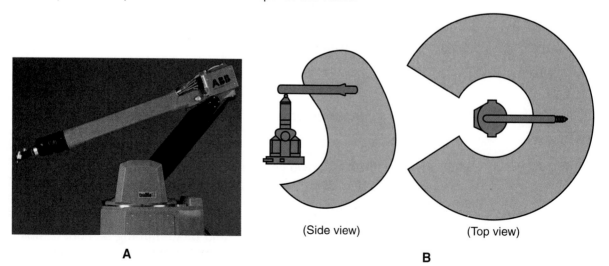

(Side view)　　　　(Top view)

A　　　　B

Figure 2-29. A—This is an example of a basic SCARA robot configuration. Note the three rotary joints and the single vertical joint used in this horizontally articulated configuration. B—This is a top view of the work envelope of a typical SCARA horizontally articulated robot configuration. This work envelope is sometimes referred to as the folded book configuration. (Adept Technology, Inc.)

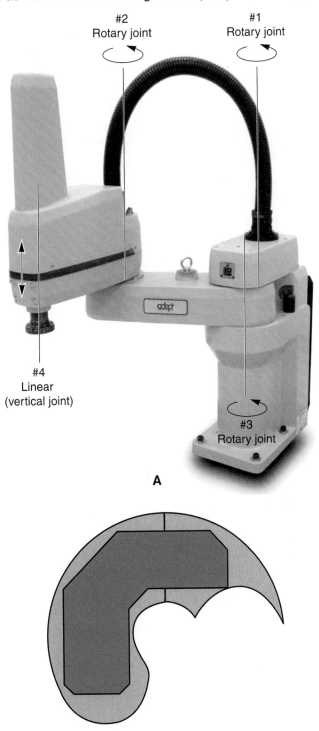

Figure 2-30. This SCARA robot is specifically designed for clean-room applications. (Adept Technology, Inc.)

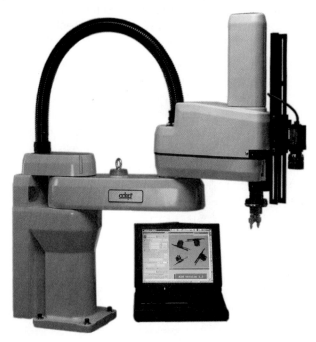

can affect accuracy, load-carrying capacity, dynamics, and the robot's ability to repeat a movement accurately. This configuration also becomes less stable as the arm approaches its maximum reach. Industrial applications of revolute configurations are discussed in more detail in Chapter 4.

Typical applications of revolute configurations include the following:

- Automatic assembly
- Parts and material handling
- Multiple-point light machining operations
- In-process inspection
- Palletizing
- Machine loading and unloading
- Machine vision
- Material cutting
- Material removal
- Thermal coating
- Paint and adhesive application
- Welding
- Die casting

Cartesian Configuration

The arm movement of a robot using the *Cartesian configuration* can be described by three intersecting perpendicular straight lines, referred to as the X, Y, and Z axes (**Figure 2-31**). Because movement can start and stop simultaneously along all three axes, motion of the tool tip is smoother. This allows the robot to move directly to its designated point, instead of following trajectories parallel to each axis, **Figure 2-32**. The rectangular work envelope of a typical Cartesian configuration is illustrated in **Figure 2-33**. (Refer to **Figure 2-13** for an example of a Cartesian gantry robot.)

One advantage of robots with a Cartesian configuration is that their totally linear movement allows for simpler controls, **Figure 2-34**. They also have a high degree of mechanical rigidity, accuracy, and repeatability. They can carry heavy loads, and this weight lifting capacity does not vary at different locations within the work envelope. As to disadvantages, Cartesian robots are generally limited in their movement to a small, rectangular work space.

Typical applications for Cartesian robots include the following:

- Assembly
- Machining operations
- Adhesive application
- Surface finishing
- Inspection
- Waterjet cutting
- Welding
- Nuclear material handling
- Robotic X-ray and neutron radiography
- Automated CNC lathe loading and operation
- Remotely operated decontamination
- Advanced munitions handling

Figure 2-31. A robot with a Cartesian configuration moves along X, Y, and Z axes. (Yamaha)

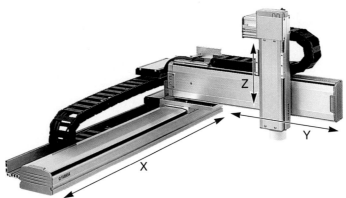

Figure 2-32. With a Cartesian configuration, the robot can move directly to a designated point, rather than moving in lines parallel to each axis. In this example, movement is along the vector connecting the point of origin and the designated point, rather than moving first along the X axis, then Y, then Z.

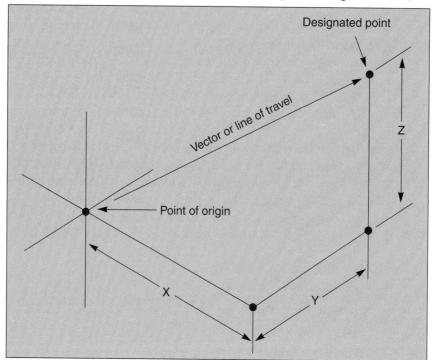

Figure 2-33. In either the standard or gantry construction, a Cartesian configuration robot creates a rectangular work envelope.

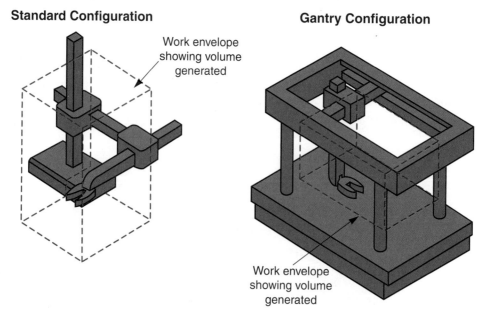

Figure 2-34. This robot has a Cartesian configuration and is used for high-precision jobs. (Adept Technology, Inc.)

Cylindrical Configuration

A *cylindrical configuration* consists of two orthogonal slides, placed at a 90° angle, mounted on a rotary axis, **Figure 2-35**. Reach is accomplished as the arm of the robot moves in and out. For vertical movement, the carriage moves up and down on a stationary post, or the post can move up and down in the base of the robot. Movement along the three axes traces points on a cylinder, **Figure 2-36**.

A cylindrical configuration generally results in a larger work envelope than a Cartesian configuration. These robots are ideally suited for pick-and-place operations. However, cylindrical configurations have some disadvantages. Their overall mechanical rigidity is reduced because robots with a rotary axis must overcome the inertia of the object when rotating. Their repeatability and accuracy is also reduced in the direction of rotary movement. The cylindrical configuration requires a more sophisticated control system than the Cartesian configuration.

Typical applications for cylindrical configurations include the following:
- Machine loading and unloading
- Investment casting
- Conveyor pallet transfers
- Foundry and forging applications

Figure 2-35. The basic configuration for a cylindrical robot includes two slides for movement up and down or in and out and is mounted on a rotary axis.

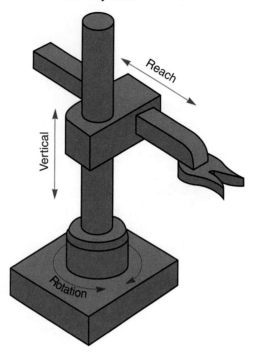

Figure 2-36. Motion along the three axes traces points on a cylinder to form the work envelope.

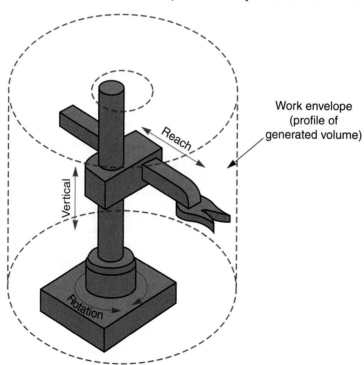

- General material handling and special payload handling and manipulation
- Meat packing
- Coating applications
- Assembly
- Injection molding
- Die casting

Spherical Configuration (Polar)

The *spherical configuration*, sometimes referred to as the polar configuration, resembles the action of the turret on a military tank. A pivot point gives the robot its vertical movement, and a telescoping boom extends and retracts to provide reach, **Figure 2-37**. Rotary movement occurs around an axis perpendicular to the base. **Figure 2-38** illustrates the work envelope profile of a typical spherical configuration robot.

The spherical configuration generally provides a larger work envelope than the Cartesian or cylindrical configurations. The design is simple and provides good weight lifting capabilities. This configuration is suited to applications where a small amount of vertical movement is adequate, such as loading and unloading a punch press. Its disadvantages include reduced mechanical rigidity and the need for a more sophisticated control system than either the Cartesian or cylindrical configurations. The same problems occur with inertia and accuracy in this configuration as they do in the cylindrical configuration. Vertical movement is limited, as well.

Figure 2-37. A pivot point enables the spherical configuration robot to move vertically. It also can rotate around a vertical axis.

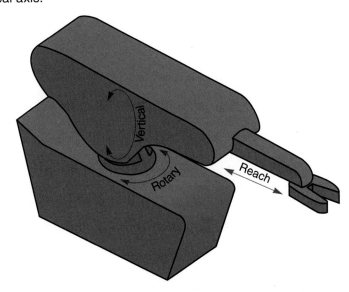

Figure 2-38. The work envelope of this robot takes the shape of a sphere.

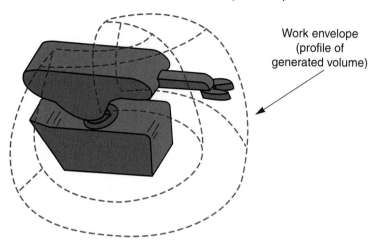

Typical applications of spherical configurations include the following:
- Die casting
- Injection molding
- Forging
- Machine tool loading
- Heat treating
- Glass handling
- Parts cleaning
- Dip coating
- Press loading
- Material transfer
- Stacking and unstacking

Special Configurations

Many industrial robots use combinations or special modifications of the four basic configurations. The robot pictured in **Figure 2-39A** uses an articulated configuration, but its base does not rotate horizontally. It is designed to literally bend over backwards in order to grasp objects behind it. This feature makes it possible to install these robots very close to other equipment, which minimizes space requirements, while maintaining a large, effective work envelope, **Figure 2-39B**. These robots are used in applications such as spot welding and material handling.

Figure 2-39. A—This heavy-duty robot literally bends over backward. (Schunk) B—The work envelope for this robot is large.

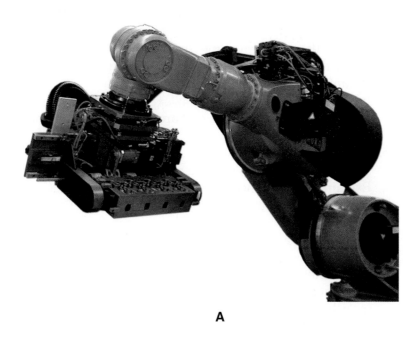

A

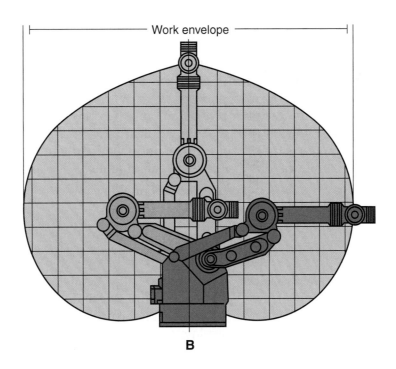

B

Review Questions

Write your answers on a separate sheet of paper. Do not write in this book.

1. Identify the five major components of a robot and explain the purpose of each.
2. What is the technical name for the robot's hand?
3. Name the three types of power supplies used to power robots. List the advantages and disadvantages of each.
4. In terms of degrees of freedom, explain why the human hand is able to accomplish movements that are more fluid and complex than a robot's gripper.
5. List and explain the six degrees of freedom used for robots.
6. Servo robots can be classified as intelligent or highly intelligent. Explain the difference between these two classifications.
7. What type of robots are considered open-loop? Explain what *open-loop* means?
8. Servo robots are considered closed-loop. Sketch a diagram of a closed-loop system and explain how it works.
9. What determines the shape of the robot's work envelope?
10. Why should you be concerned about the work envelope shape when installing a robot for a particular application?
11. What are the common work configurations used by robots? List some advantages and disadvantages of each.

Learning Extensions

1. Visit Fanuc Robotics at www.fanucrobotics.com and select videos from the side menu. This Web site provides a number of different robots performing various tasks. As you watch these video files, try to identify the robot configuration that is performing the task.
2. Visit the following robot manufacturers' Web sites. Locate the products they manufacturer and identify the robots by the configurations discussed in this chapter.
    - www.adept.com (Adept Technology, Inc.)
    - www.motoman.com (Motoman, Inc.)
    - www.apolloseiko.com (Apollo Seiko)
    - www.kawasakirobotics.com (Kawasaki Robotics (USA), Inc.)
    - www.abb.com/robots (ABB)
3. From the above Web searches, did you find any robot configuration that is manufactured more than the others. If so, identify this robot configuration and explain why you believe this to be the case.

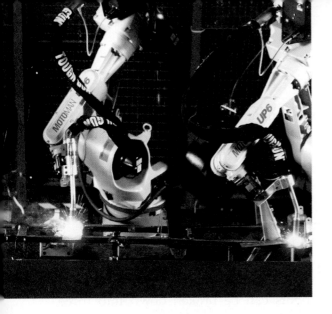

Chapter 3
Programming the Robot

Outline

3.1 The Evolution of Programming
3.2 Motion Control
3.3 Programming Methods
3.4 Programming Languages
3.5 Types of Programming
3.6 Voice Recognition

Objectives

Upon completion of this chapter, you will be able to:
- Identify the different motion control applications.
- Explain the various programming methods.
- Discuss the characteristics of the different types of programming.
- Describe various peripheral applications, such as vision and voice recognition.

Technical Terms

artificial intelligence (AI)
compiler
continuous-path (CP) motion
end stop
hierarchical control programming
high-level language
manual programming
manual rate control box
off-line programming
on-line programming
pick-and-place motion
point-to-point (PTP) motion
sensory feedback
subroutine
task-level programming
teach pendant programming
voice recognition
walk-through programming
WAVE

Overview

Robot programming has evolved along with the robots themselves. Early robots required manual settings and adjustments. Today, state-of-the-art programming is accomplished by using a computer and simple menus. The easier robots are to program, the more willing manufacturers will be to use them. This chapter covers the evolution of programming, motion control, programming methods, programming languages, types of programming, and voice recognition.

3.1 The Evolution of Programming

The evolution of industrial robots can be broken down into three periods, or generations.

- First generation—Late 1950s through the mid-1970s.
- Second generation—Mid-1970s through the mid-1980s.
- Third generation—Mid-1980s to present.

First Generation

The first-generation robots were developed between the late 1950s and the mid-1970s. They performed purely repetitive tasks and did not respond to changing conditions. Most of those found outside the laboratory were open-loop, point-to-point, and pick-and-place types. These robots often had only two or three degrees of freedom and used pneumatic or hydraulic actuators for transfer operations.

Programming these robots involved adjusting mechanical stops and limit switches. This controlled the stroke length of each programmable axis.

Careers in Robotics: Software Engineer

Software engineers design, develop, and test the computer programming and applications required to operate robotic systems and have them perform the tasks required. The programming developed must meet the specifications provided in both form (platform and language) and function (task performance). The tools software engineers use to do their job changes quickly with developing technology and specialized system applications. It is important to remain knowledgeable in the most current technology and trends in the industry.

To enter this field, most employers prefer a bachelor's degree in computer science or software engineering and knowledge of various computer systems, programming languages, and manufacturing environments. Further education and advanced degrees often open the door to advancement and more complex work.

More complex programming was done by using a rotating drum that contained actuating switches. A series of sequential moves could be set up and executed. Pneumatic or electric relay logic circuits were also commonly used. This allowed programmed steps to be controlled by a mechanical timer. The total number of programmable steps was generally between 10 and 100, depending on the robot's sophistication.

During the 1960s and 1970s, robotic research was carried out in the laboratories of industry and universities, including Stanford and MIT. A breakthrough in robot design occurred in 1975 with the introduction of robots that used computer-controlled manipulators. Microprocessors provided increased processing power at reasonable cost. This development ushered in the second generation of robot design and programming.

Second Generation

In the mid-1970s, robot manufacturers began to experiment with more advanced programming. In 1977, Unimation and Olivetti both introduced robots that could be controlled by means of programming languages. The earliest second-generation robots were not commercially successful. Some companies may have been intimidated by the complex programming and the higher cost of robots, compared to manual labor. It was not until the early 1980s that second-generation, language-programmable robots began to find acceptance in industry. They gradually began to perform such tasks as welding, spray painting, assembly, and machine loading and unloading. All of these activities required the robot to "think."

As second-generation robots became more complex, they were equipped with internal sensors and closed-loop control systems. This gave them a limited amount of feedback about their environment. Internal sensors were used to detect the robot's actual position. That position was compared to the program, enabling the robot to correct its position. Other internal sensors, such as strain gauges, were used to detect malfunctions. If a malfunction was detected, the problem was automatically corrected, or a warning was sent to the operator.

Third Generation

Third-generation robots evolved through the use of artificial intelligence. *Artificial intelligence (AI)* is the science and engineering of making machines perform operations commonly associated with intelligent human behavior. Marvin Minsky, one of the fathers of AI, defines it as "the science of making machines do things that would require intelligence if done by men." Scientists and engineers have been studying AI since the 1930s. Practical use, however, was largely confined to laboratory research until the late 1980s.

Third-generation robots are capable of sensing their environment and making intelligent decisions to perform tasks more efficiently. Sensing devices, such as vision sensors, provide robots with information about their surroundings, **Figure 3-1**. Robots use this information to determine how to proceed in performing a task. For example, vision systems are used for part recognition allowing robots to distinguish between good and bad parts on a production line.

Figure 3-1. Vision guidance tracks parts on a moving conveyor belt. (Adept Technology, Inc.)

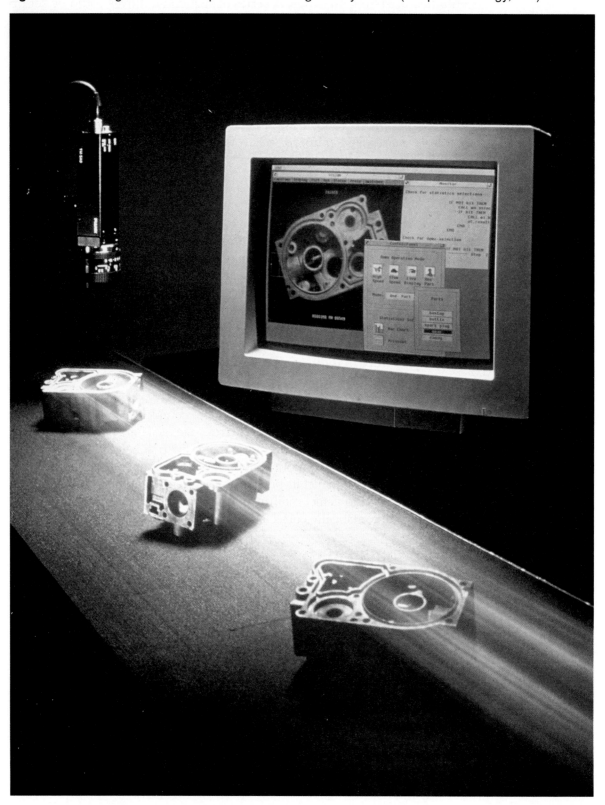

3.2 Motion Control

A robot's manipulator moves through a series of points while performing a task. Robots can be classified according to their pattern of motion. The three classifications are pick-and-place motion, point-to-point motion, and continuous-path motion.

Pick-and-Place Motion

Limited-sequence robots use *pick-and-place motion* to move the end effector to the correct position. Pick-and-place motion is often used in manufacturing processes to perform work that is repetitive and does not require many complicated movements to accomplish a task, such as picking up a part at one location and placing it in another location.

One method of programming the pick-and-place pattern of motion is to manually set mechanical stops or limit switches for each designated point. This method allows for two positions per axis. Programming additional positions per axis requires extra controls and additional time spent in the process. Because of the laborious programming, the number of points the robot can move through is comparatively low.

Figure 3-2. Pick-and-place motion involves only two positions per axis.

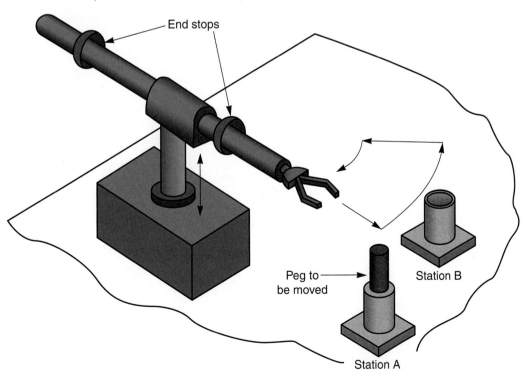

The end effector follows a fixed pattern of movement, as illustrated in **Figure 3-2**. Generally, only one axis of the robot moves at a time. *End stops* are devices used to control the length of travel along an axis. Once in place, end stops prevent movement past a certain point. Note that the position points used by pick-and-place robots are points along the various axes, not points in space.

Point-to-Point Motion

Point-to-point (PTP) motion involves the movement of a robot through a number of points in space. The programmer uses a combination of manipulator axes to position the end effector at a desired spot. The positions are recorded and stored in memory. During playback, the robot steps through the recorded points. The path of motion is a series of straight lines between the points, **Figure 3-3**, with no real control of the end effector's path *between* the taught points. As in pick-and-place motion, point location is more important than controlling the path of travel.

In order to better understand point-to-point motion, review the motion illustrated in **Figure 3-4**. Suppose the task to be taught is to take a peg out of its holder and insert it into a holder in a different location. However, instead of using end stops to control length of travel, the actual locations are recorded into the robot's memory.

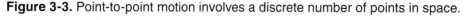

Figure 3-3. Point-to-point motion involves a discrete number of points in space.

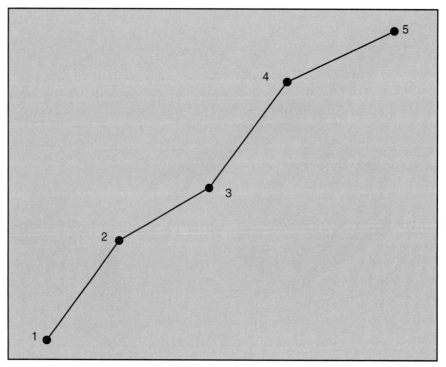

Figure 3-4. Illustration of the steps involved in the point-to-point operation example.

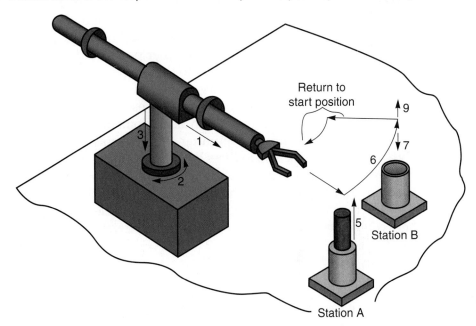

The peg is initially located in a holder at station A. The robot's arm is retracted and the gripper is open. The point of insertion is at station B. The steps to be recorded are as follows:

1. Move the arm until the gripper is located above the peg. Record this point into memory.
2. Adjust the wrist joint of the robot until the gripper is properly aligned for grasping the peg. Record this position.
3. Move the gripper down over the peg to the point where the gripper will be able to grasp the peg when closed. Realign the gripper with the peg by adjusting the various joints. Record this point.
4. Close the gripper on the peg and record the point.
5. Carefully lift the peg from the hole vertically. After the peg is out of the hole and is at the desired elevation, record that point.
6. Move the arm until it is approximately over the center of the hole at station B. Record this point.
7. Carefully lower the peg, adjusting the various joints until the peg is properly inserted into the hole. Record this point.
8. Open the gripper to release the peg. Record the point.
9. Move the arm until the gripper is located at some point directly above the peg and record this point.
10. Stop the robot. Move the peg back to the hole at station A.

After the points have been recorded, take the robot out of teach mode. The robot returns to the start position and steps through the sequence using the various points recorded into memory. The robot in this example would have to be stopped after step 9 because no additional pegs were placed at station A. In a real work situation, another peg would appear at station A or additional steps would be added to the program.

During playback, the trajectory of the end effector is generally different from the path used by the operator when establishing the points. This is because the operator must move each axis or joint independently. However, if a joystick is used during programming, the robot will move along a straighter path. Using a joystick also reduces the programming time involved.

Several stops along a given axis can be programmed. Point-to-point servo robots are capable of storing hundreds of discrete points in space, far more than the two stops in pick-and-place motion. Acceleration and deceleration between points is controlled by a separate device, such as a tachometer.

Continuous-Path Motion

Continuous-path (CP) motion is an extension of point-to-point motion. The difference is that continuous-path motion can involve several thousand points. Since more points are used, the distance between each point can be extremely close. Using a great number of points results in movement that is smooth and continuous, **Figure 3-5**. With continuous-path motion, control

Figure 3-5. Continuous-path motion involves an infinite number of points.

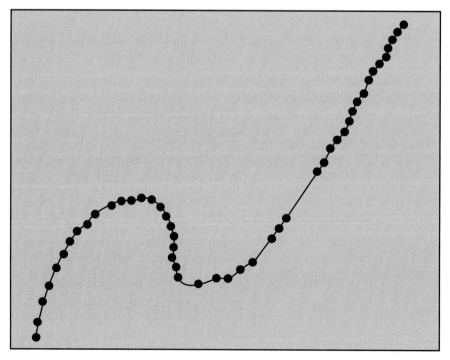

of the end effector's path is more important than end point positioning. The robot generally does not come to rest at various points, as is often required in PTP motion.

Programming is done by an operator who physically moves the end effector through its motions. The positions on the various axes are recorded. Some continuous-path robots record up to 80 points per second.

Playback rate can be changed to provide the best operating speed for the task. For certain applications, such as spraying, it may be better to program the robot at a slow speed and play the program back faster. Other applications, such as arc welding, require faster programming than playback.

Continuous-path motion offers several advantages:

- Movement is smooth and continuous.
- Programming is simple and no prior knowledge of programming is required.
- Velocity and acceleration of the end effector can often be controlled.
- The operator must simply understand the operation he or she is trying to teach the robot to be successful in programming it.

There are, however, disadvantages to continuous-path motion:

- A large amount of memory is required to store all the points recorded for a task. In the past, memory was very expensive. Memory chips are now more reasonably priced, so this is not as significant a disadvantage as it once was.
- Since recording during programming is constant, undesirable motions are recorded into memory in addition to intended movements. The robot duplicates all the movements recorded during playback.
- The robot's arm must be counterbalanced and able to move freely without power so the operator can produce smooth-flowing motion while programming.

3.3 Programming Methods

Robots can be programmed manually, by means of a teach pendant, by walking them through a task, or by means of a computer terminal. The various programming methods are best suited for only particular types of motion control.

Manual Programming

Robots with pick-and-place, point-to-point, or open-loop controllers can be manually programmed. *Manual programming* can be best described as a type of machine setup. An operator adjusts the necessary end stops, switches, cams, electric wires, or hoses to set up the sequence, **Figure 3-6**. This type of programming is typical of first-generation limited-sequence or pick-and-place robots. Even though these robots appear to be simple in nature, they are capable of performing many manufacturing tasks. If an

Figure 3-6. In manual programming, the mechanical stops and limit switches are manually adjusted to control the robot's movement. (ABB Robotics)

application is suited for a less sophisticated robot, there is no reason to invest in a more complex and costly model.

Manual programming is typically simple and does not require an operator skilled in the use of computers. The capital investment and maintenance costs for manually programmed robots are low. These robots are capable of high operating speeds and have good accuracy and repeatability. However, the flexibility of manually programmed robots is limited, and they may have only two or three degrees of freedom. Control of intermediate points along the path is generally not available. Only two positions are programmed for each axis and, depending on the complexity of the robot, the total number of programmable steps is generally 10 to 100.

Using a Teach Pendant

In *teach pendant programming*, the operator leads the robot through the various positions involved in an operation, **Figure 3-7**. As the end effector reaches a desired point, that point is recorded into memory by pushing buttons on the teach pendant. The recorded points are used to generate a point-to-point path the robot follows during operation.

Figure 3-7. The operator is using a teach pendant to program the robot through various moves. (ABB Robotics)

The teach pendant is a popular method of programming because it is convenient, simple to learn, and suitable for programming many tasks found in industry. This method does not require an operator skilled in the use of computers. However, complex motions and applications that require close tolerances may involve a lengthy programming time. The robot must be operating while programming, as the program cannot be entered into the teach pendant while the robot is off-line.

Walk-Through Programming

Walk-through programming is used for continuous-path robots. An experienced operator physically moves the end effector through the desired motions. While the robot moves along the desired path, as many as several thousand points are recorded into memory. The number of points recorded can vary from one manufacturer to another. As the number of points sampled per second is increased, the movement of the robot becomes smoother and more fluid. However, as the rate of sampling increases, so does the need for greater storage capacity. Spraying and arc welding are the most common applications programmed using the walk-through method. Other applications include grinding, deburring, polishing, and palletizing.

Walk-through programming does not require computer experience. However, the person programming the robot must be highly skilled in the precise motion required by the task. The robot must be operating while programming using the walk-through method.

Using a Computer Terminal

Programming a robot using a computer can be done *on-line* (at the robot's console) or *off-line* (away from the robot), **Figure 3-8**. Final testing of the program is done at the job site.

Computer programming provides greater flexibility. It is not necessary to take a robot out of operation while the program is being written and debugged, so productivity is not affected. High-level computer languages allow programming of more complex operations. These languages reduce programming time, which increases productivity. The primary disadvantage is that the operator must be experienced in the use of computers, high-level languages, and programming logic.

3.4 Programming Languages

The first robot programming language, known as *WAVE*, was developed at the Stanford Artificial Intelligence Laboratory in 1973 for research purposes. Most manufacturers that provide off-line programming have developed their own language, as there is not an established set of standards in the industry. Some common languages and their sources are listed in **Figure 3-9**.

Figure 3-8. A personal computer used to program a robot off-line. (ABB Robotics)

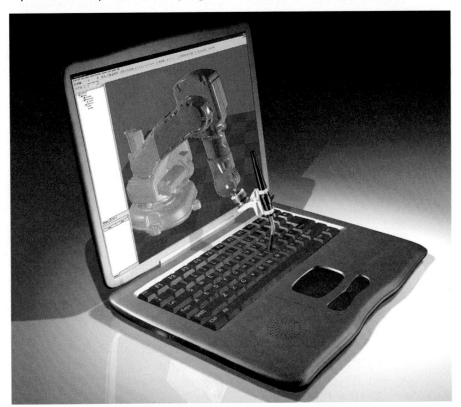

Figure 3-9. Some common robot programming languages and corresponding originators.

Common Programming Languages		
Robotics Programming Language	**Year of Development**	**Originator**
SAIL (Stanford Artificial Intelligence Language)	1968	Stanford University
AL (Assembly Language)	1970s	Stanford University
MCL (Manufacturing Control Language)	1980	McDonnell Douglas Corp.
VAL (Variable Assembly Language)	1980	Unimation Inc.
Karel	1981	Fanuc Robotics America Inc.
AML (A Manufacturing Language)	1982	IBM
RAIL	1982	Automatrix
RPL	1984	Hewlett Packard
RobotBASIC	1984	Intelledex Inc.
Magik	1999	GE Energy

High-level languages are programming languages that more closely resemble standard English. Because computers cannot understand English, these languages are translated into machine code by means of a program called a *compiler*.

Even though robot programming languages have been developed by different groups, many are similar. The main difference is in the choice of key words and commands. Each language also has its own syntax, or structure. MELFA BASIC IV, AML, and Karel, for example, are computer programming languages related to BASIC, Pascal, and PL/I languages, with instructions given in the form of subroutines. A *subroutine* is a set of instructions within the program that has a beginning and an end.

Each language defines the number of characters allowed for a program name and use of special characters. With MELFA BASIC IV, for example, the program name must start with a letter and may contain up to twelve characters (letters and numbers). Also, comments may be included within the program code by preceding the comment text with an apostrophe ('). Unlike many programming languages, MELFA BASIC IV does not require a character to signal the end of a line of code. The following sample of MELFA BASIC IV program code provides instructions to move the robot's end effector from location A to location B.

```
10 'Program Name: Program Exe_2
20 'Author: User
30 'PSAFE is a position clear of all parts and obstructions
40 'P1 is a XYZ location to be taught
50 'P2 is a XYZ location to be taught
60 MOV PSAFE
70 MOV P1 DLY 1.0
80 MVS P2 DLY 1.0
90 MOV PSAFE
100 END
```

With Karel, the program name can contain up to twelve characters, including letters, numbers, and underscores. The program name must start with a letter. Comments can be included within the code by preceding them with two hyphens (--). An identifier follows key words. A semicolon (;) is not required at the end of each line.

The following is a program written in Karel to move the robot's end effector from location A to location B.

1. PROGRAM Exe_2

2. Variables

3. A: Position --Position location to be taught

4. B: Position --Position location to be taught

5. --Line left blank for separation--

6. Begin --Program execution

7. Move to A

8. Move to B

9. END Exe_2

3.5 Types of Programming

The majority of robots in industry today use hierarchical control programming. However, task-level programming simplifies the programming task and is growing in popularity.

Hierarchical Control Programming

As discussed in Chapter 2, hierarchical control separates program control into a number of different levels. In *hierarchical control programming*, each level accepts commands from the level above and responds by generating simplified commands for the level below, **Figure 3-10**. This system uses *sensory feedback* (input from the environment) to close control loops. In other words, sensor inputs affect how the robot responds.

Figure 3-10. This diagram shows a hierarchical control structure.

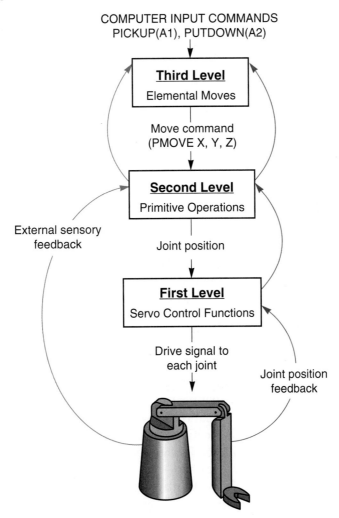

First Level

Servo control functions are computed at the lowest level in the hierarchy. This is the level at which most robots in use before 1980 were programmed and controlled. Program commands are compared with feedback from position indicators. If the values are different, a drive signal is generated to move each joint until the error signal is zero.

Programming at the first level does not require the use of a computer. The arm is moved using a *manual rate control box*. The box consists of a knob and some switches that control the movement of each axis individually. (A rate control box should not be confused with a teach pendant, as it does not have the same capabilities.) The joints are moved one at a time until the robot arm is in the desired location. The values of the position indicators are then all stored in memory. The arm is moved to another point, and the process is repeated until the desired path is stored in memory. This form of programming is time-consuming and tedious.

Second Level

A computer is required to program a robot at the second and third levels. These levels require real-time interaction of the robot with its environment by means of sensors. Commands are issued for primitive operations, such as a move along a straight line defined by the vector X, Y, Z. These commands are translated by the computer into the proper position values.

The second level in the hierarchy receives position feedback and generates the command sequence used by the first control level to accomplish the task. At this level, a joystick can be used for programming instead of the manual rate control box. The operator does not have to worry about moving individual joints. Programming tasks are much easier and faster.

Third Level

The third level in the hierarchy receives commands for elemental moves. A typical command is MOV P1 or MOVE to A. When these commands are issued by third-level circuitry, the control system monitors feedback signals. It then generates a sequence of commands for primitive operations. These primitive operations become the input to the next level down.

For example, to program a simple transfer task, commands for two elemental moves are typed in the desired sequence into the computer. In the MELFA BASIC IV language this would be:

MOV PSAFE (Safe location away from obstructions)
MOV P1 (X, Y, Z location above position A)
MVS P2 (X, Y, Z location to pick up part)
HCLOSE (Close gripper)
MVS P1 (X, Y, Z location above position A)
MOV P3 (X, Y, Z location above position B to place part)
MVS P4 (X, Y, Z location to place part)
HOPEN (Open gripper)
MOV P3 (X, Y, Z location above placed part)
MOV PSAFE (Safe location away from obstructions)

This set of commands causes the robot to move from a safe location to location X, Y, Z (position two) and pick up an object. It then moves the object through a series of points to location X, Y, Z (position four) and puts the object down. The same sequence is repeated for the next set of positions. These points are previously recorded using a joystick, teach pendant, or other device, moving the robot through the points manually. The X, Y, and Z locations can also be typed directly into memory.

During playback, the robot grasps and releases at the recorded locations. The environment must be tightly controlled. The locations of the objects to be grasped and their orientation must always be the same. A slight misalignment can cause damage to both the robot and workpiece. Controlling the environment often requires additional machinery and equipment that can cost several times more than the basic robot.

Programming even a simple task using hierarchical control and traditional high-level languages can take days. Additional time must be set aside for testing and making changes. Even after the system is up and running, further changes may be required.

Task-Level Programming

In today's manufacturing environment, fast production time and low system costs are critical to a company's success. Yet today's robot systems are expensive and difficult to install and maintain. A typical robot in an automated work cell might move something, assemble parts, use vision for guidance and parts inspection, control local work cell devices, respond to sensor inputs, keep statistics on processes, and communicate with a host computer or an operator.

Figure 3-11A shows the architecture for a typical custom work cell. It is more complex than it needs to be. Four separate controllers are used. A motion control system runs the robot, while the programmable logic controller (PLC) provides work cell logic. The vision system is for inspection and robot guidance. The personal computer is included for user interface. What is needed is a simpler system with user-friendly hardware and software. The answer lies in task-level programming.

In *task-level programming*, the user specifies the goals of each task rather than the motions required to achieve those goals. Instructions are entered using simple English-like terms. The details for each and every action the robot is to perform do not need to be specified. This allows the use of instructions at a significantly higher level than those produced by languages such as MELFA BASIC IV, AML, Karel, or RobotBASIC. Many activities are programmed automatically by the computer.

An example of a task-level programming environment is shown in **Figure 3-11B**. In this example, only one controller and one programming system are needed to work with multiple technologies. The need for programming knowledge is eliminated by straightforward menus and instructions entered in English. The operator can focus attention on the task instead of writing programming code.

Figure 3-11. Using task-level programming streamlines the programming process. A—This is a typical custom work cell layout. The arrows indicate interactions. B—A single controller and programming system are used to work with multiple technologies. (Adept Technology, Inc.)

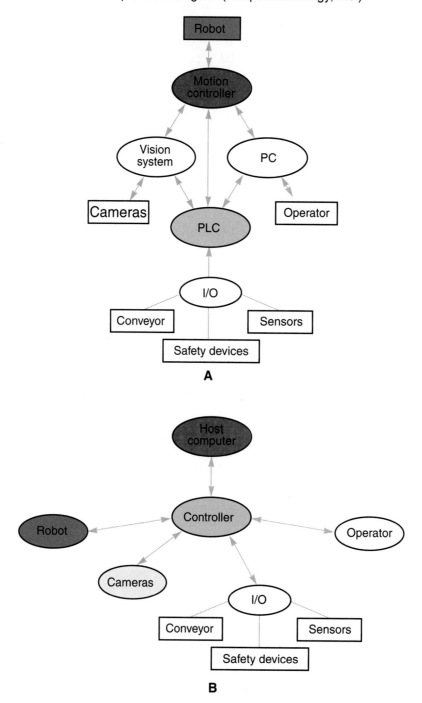

Interactive screens are used to create, teach, debug, and run application programs. Programming software has become so user-friendly, even inexperienced programmers can program a robot. Some common task-level programming functions are described in **Figure 3-12**.

Task-level programming is a highly efficient environment that replaces hundreds, even thousands, of lines of programming code with a small number of menu-selected statements. Representative programming screens are shown in **Figure 3-13**.

3.6 Voice Recognition

Voice recognition or speech recognition systems use recognizable words or phrases as a form of audio data entry, rather than using a keyboard. Utilizing voice recognition can produce the same result as typing words on a keyboard. Only a few years ago, voice recognition was experimental and existed only in the lab. In today's society, however, voice recognition systems are commonplace. Many organizations use voice recognition with their telephone answering systems. Rather than

Figure 3-12. The basic program functions and steps are defined on-screen by the programming software. Programmers use on-screen sequence-editing functions to develop the robot's movement.

\multicolumn{2}{c}{**Common Task-level Programming Functions**}	
Function	**Programming**
Robot Movements	The first step in creating the robot program is to select the sequence of movements from a menu. For example, the MOVE statement tells the robot to pick up parts from one location and place them in another location. This type of sequence may be used to move a part from a conveyor belt to a pallet.
Location Information	A location database stores work cell locations. The height of each location, approach, and departure are taught or modified by making the appropriate selections and entries. The robot's speed, type of motion, and other details are also selected and stored.
Palletizing	The PALLET statement is commonly used in automatic palletizing functions. The operator can define spacing and the number of pallet locations.
Visual Sensors	The type of visual sensors needed can be specified by making the appropriate selections and entries on the program interface. For example, a visual inspection sensor can locate and evaluate parts, while a visual guidance sensor may track and orient the parts along a conveyor belt.
General Control Functions	From the control panel screen, the operator can start the operation, slow the speed of the robot, and step the robot through its motions to be sure the program is performing as intended. The control panel screen may also be used for debugging and cell control.

Figure 3-13. Representative programming screens. A—HexSight by Adept is a machine vision software that is used to locate and inspect parts. B—Vision system going through a calibration cycle. C—Vision system performing an inspection on an IC footprint. (Adept Technology, Inc.)

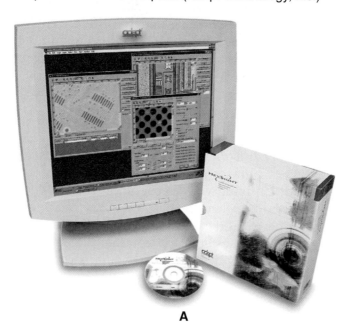

A

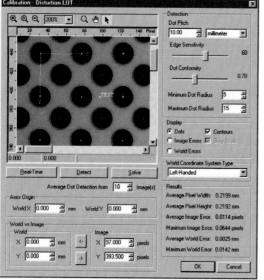

B

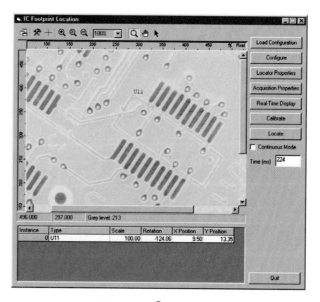

C

using a telephone keypad to enter a number selection, you are given the option to speak your selection.

Voice recognition has improved dramatically over the past few years. Robots have been developed that understand basic conversation phrases in multiple languages and can respond in the appropriate language. Voice recognition is being considered in the manufacture of items for physically handicapped people. Robots equipped with two-way communication can meet the needs of a person who is completely paralyzed. In this situation, using the person's voice may be the only practical means of communication. Additionally, experts suggest that voice frequency is just as accurate as fingerprints for identification purposes.

Voice communication from a robot may also serve to diagnose and warn of possible problems in a manufacturing operation. If the robot is experiencing trouble with one of its own components, the problem could be verbally communicated to an operator or service person.

One of the biggest hindrances in mainstreaming this technology is likely to be the cost. However, certain key advantages make voice recognition attractive. Verbal communication is inexpensive and voice recognition can reduce set-up time. This is important since manufacturing trends are shifting away from volume and toward variety. Operators may be intimidated by some of the other set-up methods, but with voice recognition, anxiety is reduced. Voice recognition can truly be described as user-friendly.

Review Questions

Write your answers on a separate sheet of paper. Do not write in this book.

1. The evolution of programming can be broken down into three periods. List these and discuss the developments in each period.
2. There are three classifications for robotic motion control. List these classifications and identify the differences between them.
3. Assume that you are teaching paint spraying to a robot that uses continuous-path motion. What might happen if you fail to specify each necessary point while in the teach mode?
4. Identify four methods used to program robots. Briefly describe each method and provide an application for each.
5. How are pick-and-place robots programmed?
6. What is the advantage of using a personal computer to create robot programs?
7. What are some advantages of task-level programming?
8. Explain how programming with integrated voice recognition would be useful with an industrial robot.
9. What method is used to convert high-level programming languages to machine code that the robot's microprocessor can understand?
10. Most companies use proprietary programming languages for the robots they manufacture. What is the disadvantage of this approach for the end user?

Learning Extensions

1. Using the Internet, conduct a search to find four robot manufacturers and identify the programming language they use to program the robots they manufacture.
2. Conduct a web search on Honda's ASIMO and view the various Web sites. As you surf these sites, examine the history and evolution of humanoid robot construction.
3. Visit the official Honda Web site and select ASIMO. View the various videos of the functions of ASIMO. Identify possible uses for this type of robot.

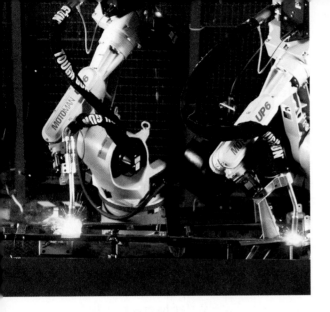

Chapter 4
Industrial Applications

Outline

4.1 Integrating Robots into the Manufacturing Process
4.2 Selecting the Right Robot
4.3 Using Robots in Industry

Objectives

Upon completion of this chapter, you will be able to:
- Describe how robots are integrated into a manufacturing process.
- Select the proper robot for a given task.
- Identify processes where robots are used.
- List peripheral devices used to complete tasks.

Technical Terms

accuracy
automated guided vehicle (AGV)
command resolution
design for manufacturability
dynamic performance
fixturing
interlocks
light curtain
operational speed
payload
pressure sensitive safety mat
repeatability
resolution
service robot
spatial resolution

Overview

This chapter discusses applications of industrial robots, beginning with a section on integrating robots into the manufacturing process. This includes design for manufacturability and the technical factors considered when selecting robots. Additionally, safety considerations and guidelines and typical uses for robots in industry are covered.

81

4.1 Integrating Robots into the Manufacturing Process

Automation is a high priority for manufacturing companies worldwide. Competition, improved technology, cost, market conditions, productivity, available manpower, and undesirable work environments are forcing manufacturers to install a greater number of robots. How can robots be successfully incorporated into a manufacturing process? Of the many factors to consider, it is common to start with the design of the manufactured products and safety.

Design for Manufacturability

Design for manufacturability means designing products with the robots that will assemble them in mind. First, the engineering department must decide which parts of the product can be assembled by robots and which need human assembly or dedicated equipment. The next step is to design, or redesign, the parts of the product for ease of robotic assembly. The list of design tips that follows is useful to consider when designing a product for robot assembly. The following tips will help design a robot that works efficiently and moves to designated positions accurately and consistently.

- Minimize the number of parts to reduce complexity, **Figure 4-1**.
- Reduce the directions of approach required to assemble a product. This allows a less sophisticated robot to be used.
- Minimize the number of obstructions in the environment so that the robot can work in a straight line.
- Whenever possible, use subassemblies with components that stack on top of each other or that can be assembled in sequence using downward motion.
- Add chamfers, guide pins, ridges, and other physical characteristics that allow the robot to lock the part into its proper location. The cost of these guides is far less than the cost of robot vision systems.
- Simplify fastening by using tabs, snaps, or other methods that allow parts to be joined in one motion, **Figure 4-2**.
- Eliminate screws, springs, and adjustments.
- Avoid compressible parts that robots do not handle well and can deform, such as wires, foils, or foams.
- Give parts common features. For example, all parts might have the same size hole or post in the center. This allows the same end effector to pick up any of the parts.
- Make parts as symmetrical as possible to ensure correct assembly, no matter how the robot picks them up, **Figure 4-3**. If a part must be asymmetrical, use the asymmetrical features to assist in orientation.

Figure 4-1. A—This subassembly contains a total of eight separate parts, making robotic assembly very difficult. B—The same part is redesigned to accommodate robotic assembly. The brackets are punched out of the original base plate, resulting in an 8-to-1 parts reduction.

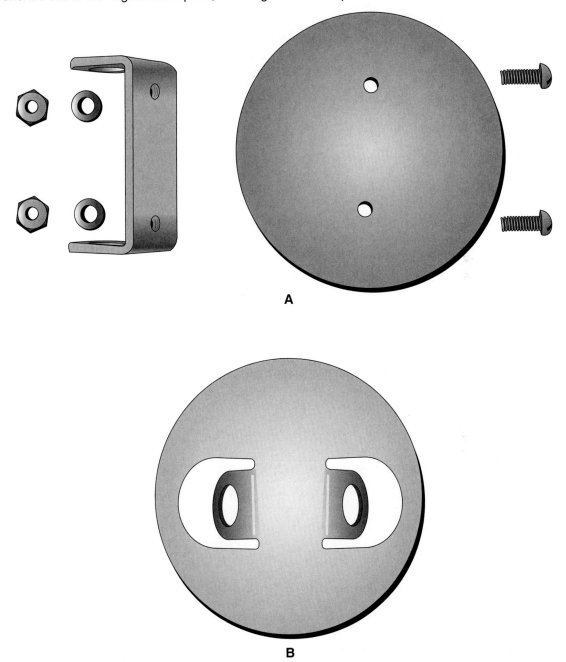

Many companies have discovered that when they design for manufacturability, the process becomes so simplified that it is faster and more productive to use human workers or dedicated machines than to incorporate robots. If demand is sufficient, all three methods might be used.

Figure 4-2. These screws have been designed specifically for ease of assembly, with features such as self-tapping, self-threading, and a washer attached to the screw head.

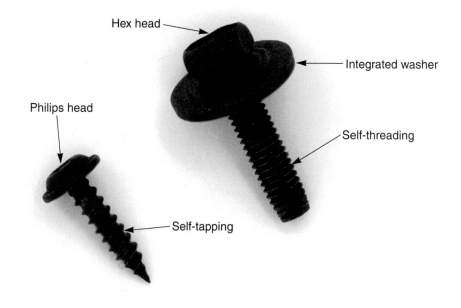

Figure 4-3. The base plate on the right contains asymmetrical holes, which means that it must be installed using one particular orientation. The redesigned base plate on the left is symmetrical and does not require any particular orientation during assembly.

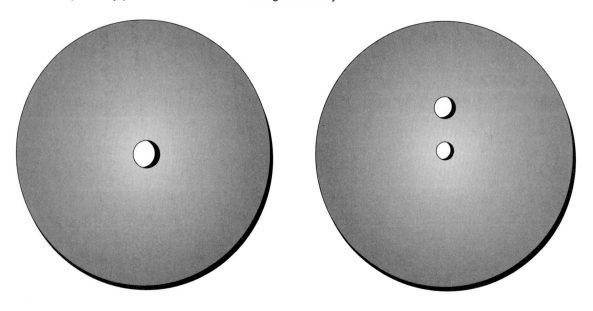

Robotic Safety Considerations

An important consideration in installing robot systems is *safety*. Robots differ from other machinery due to their degrees of freedom and sizeable work envelopes. Providing a safe working environment must be considered during the design, installation, maintenance, and operation of robot systems. A robotic work station can be dangerous if workers are not properly trained or not made aware of possible hazards.

The U.S. Department of Labor: Occupational Safety & Health Administration offers information and safety guidelines for the safe operation of robots. "Industrial Robots and Robot System Safety" can be found in the *OSHA Technical Manual*. Topics included in the information presented are robot classifications, typical hazards, and safeguarding personnel.

The American National Standards Institute published safety requirements related to industrial robots in *Industrial Robots and Robot Systems—Safety Requirements* (ANSI/RIA R15.06-1999). This document provides safety requirements for industrial robot manufacturing environments, installation sites, and general operation and personnel safety.

General Safety Guidelines

Safety considerations should begin with research into the characteristics of the specific robot(s) being used.

- What is the size and shape of the robot's work envelope?
- What methods of motion control are used?
- What are the limits for a safe payload?
- What is the range of operating speeds?
- What other features are provided by the manufacturer?

The design of the work station should reflect safety. Omit pinch points by being aware of fixed objects within the work area, such as vertical poles. Keep the end effector in mind when designing the space. The relationship of the end effector to work fixtures and other equipment must be taken into account. Outline the full range of the end effector's motion on the floor using paint or colored tape. This provides a constant, highly-visible perimeter of the robot's work area. The work station should also include barriers that cannot be passed through or over. The design should place control panels outside the work envelope and provide interlock access to the robot controller. *Interlocks* are safety devices that are designed to prevent unauthorized access to hazardous areas by requiring a key for entry. Implementing two-step procedures to resume operation after a shutdown is an additional precaution to ensure worker safety. The entire design should be documented, including any changes made during installation and debugging. This is important information to purchasing, operating, and maintenance personnel.

Employee training and a conscientious effort to provide safe working conditions are important components of every safety program. Operating personnel, programmers, maintenance workers, and clean-up crews need extensive job- and robot-specific training. Periodic refresher courses should be scheduled for as long as robots are in use.

Safety Barriers and Sensors

Safety measures for a robot are more complex than for other kinds of equipment. Two primary safety devices used are barriers and sensing systems.

One of the safest barriers is a fence, **Figure 4-4**. A safety fence prevents unauthorized entry and also safely contains dislodged work pieces. A

Figure 4-4. A steel safety fence surrounds this arc welding robot. The control for the robot is mounted outside the fence. The robot cannot be started until the operator steps outside and closes the gate.

safety fence is at least 6′ high and constructed of wire mesh, safety glass, or rigid plastic sheets, **Figure 4-5**. In toxic applications, such as paint spraying, a solid enclosure will also protect personnel from chemicals. However, fencing makes it more difficult to move or rearrange the robot and other work cell equipment. Fences take up valuable floor space and often obstruct the view of critical operations. Fences may also be an obstruction when trying to teach the robot or perform preventive maintenance.

Another kind of barrier is the *light curtain*, which consists of photoelectric, presence-sensing devices (**Figure 4-6**). If a worker enters the work envelope, the sensors send a signal to cut the power to the robot. An infrared light curtain is programmable, which allows certain areas to be excluded from the sensing devices. This permits access to equipment that is not within the reach of the end effector. One drawback of the light curtain is that moving equipment can trip the system.

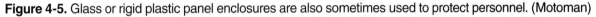

Figure 4-5. Glass or rigid plastic panel enclosures are also sometimes used to protect personnel. (Motoman)

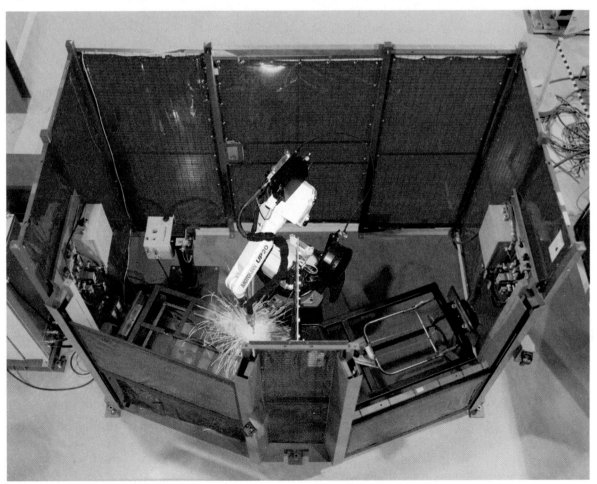

Figure 4-6. This equipment projects a curtain of light. When the light is interrupted, power to the robotic equipment is shut down. (Motoman)

One of the simplest sensing systems is a *pressure sensitive safety mat*. Sensors embedded in the mat send a signal to the controller when weight is placed anywhere on the mat. The controller then sends a "shut down" signal to the connected equipment. Pressure sensitive safety mats can be easily placed in dangerous areas within a facility and will automatically turn off dangerous machinery when a worker steps on the carpet.

Some sensing systems make use of a combination of capacitance (transfer of an electrical charge from one conductor to another), infrared, ultrasonic, or microwave sensors to detect a worker's presence and track them within the work envelope. The power is cut off whenever a worker enters the robot's work envelope.

4.2 Selecting a Suitable Robot

With hundreds of robot models available today, selecting the best robot for the job can be a major task. Robots occupy a position between human workers and equipment designed to perform one specialized function. Robots can perform repetitive tasks without tiring and can operate in hostile environments. They can maintain greater precision than human workers when performing repetitive work.

However, there are two main areas for which robots are not suited. They should not be used for long, high-speed runs producing many identical items. Dedicated machinery is better suited for that type of work. Conversely, robots should not be used for short, complex tasks that require a high degree of hand-eye coordination. Currently, these jobs are still best suited to human workers. Otherwise, robots are useful for many production jobs. In addition to environmental conditions and coordination with associated equipment, other important considerations when selecting a robot include the work envelope, degrees of freedom, and other measures of performance.

Work Envelope

As discussed in Chapter 2, the work envelope is different for each robot configuration and is defined by its type of joints and degrees of freedom. All of the fixtures and other machinery that will work with the robot must be located within the work envelope. Many times, placement of support equipment dictates the shape of the work envelope. To provide greater flexibility, some manufacturers offer optional wall and ceiling mounts in addition to the robot's standard floor mounting.

Degrees of Freedom

The number of degrees of freedom determines the robot's ability to move within the work envelope. Robots with six degrees of freedom can work anywhere within their work envelope. However, not all applications require such a versatile robot. The configuration of each individual work station determines the requirements.

Sometimes, an additional degree of freedom can be obtained by mounting the robot on rails. This permits it to move horizontally, increasing its overall reach. The six possible degrees of freedom can be broken into two basic categories: body and wrist. The table in **Figure 4-7** provides descriptions of the six degrees of freedom and the associated axes.

Resolution

Resolution is determined by the robot's control system and refers to the smallest incremental movement the robot can make. Resolution can also be described as the smallest segment into which the work space can be divided. *Command resolution* is the closest distance

Figure 4-7. This table summarizes the six degrees of freedom and the corresponding robot axes.

Robot Axes and the Six Degrees of Freedom		
Degree of Freedom	Robot Axis	Description
First	Waist (rotational transverse)	Movement about a vertical axis.
Second	Shoulder (radial transverse)	Extensions and retraction of the arm.
Third	Elbow (vertical transverse)	Up and down motion.
Fourth	Pitch	Up and down movement of the wrist.
Fifth	Yaw	Side to side of the wrist.
Sixth	Roll	Rotation of the wrist.

between movements and is calculated by dividing the travel distance of each joint by the number of control increments. For example, a robot travels a distance of 36″ (91.44 cm) along one axis and the controller is programmed with a total of 8000 points. Divide 36 by 8000 to produce a command resolution of 0.005″ (0.0127 cm), without taking into account any mechanical inaccuracies.

The manipulator uses mechanical components to position the end effector at the various points in the work envelope. Whenever mechanical members are used (gears, chains, cables, or ball lead screws), certain inaccuracies occur—chains and cables stretch, and gears are prone to backlash. Also, overweight payloads can create certain inaccuracies. All these factors can affect resolution.

Tool positioning during programming is an important factor in accomplishing the desired task. However, tool position may vary due to the command resolution and mechanical inaccuracies. *Spatial resolution* describes the accuracy of movement of the robot's tool tip. Spatial resolution takes into account command resolution and mechanical inaccuracy. For example, if the command resolution of 0.005″ (0.0127 cm) has a mechanical inaccuracy of 0.003″ (0.0076 cm), then the spatial resolution would be 0.008″ (0.0203 cm). This figure is derived by adding the command resolution to the mechanical inaccuracy.

In the work envelope, spatial resolution varies with tool location. The consistency of the spatial resolution value within a work envelope is affected more by certain robot configurations than by others. For example, the Cartesian configuration offers fairly constant spatial resolution. Rotary joints, however, can impact the spatial resolution value at various points within the work envelope, resulting in necessary adjustments.

Accuracy

Accuracy and spatial resolution are both by-products of command resolution and mechanical inaccuracies. *Accuracy* expresses how precisely the robot's hand is programmed to reach a predetermined point. The value associated with accuracy is expressed as half of the spatial resolution. A robot with a spatial resolution of 0.010" (0.0254 cm) has an accuracy of 0.005" (0.0127 cm).

Large robots with payloads of 100 lbs. (45.36 kg.) or more have an average accuracy of 0.050" (0.127 cm). Small robots used for tasks where payloads are lighter, such as assembly operations, have an average accuracy of 0.002" (0.005 cm). Improved accuracy increases with advancements in technology.

The speed of movement and weight of the payload also affect accuracy. As speed increases, accuracy decreases. To stop the manipulator at a given location, speed must be reduced to prevent overshooting the position. If the speed is reduced too much, however, valuable time is wasted. Overcoming the inertia in moving a heavy load can also affect position accuracy. To overcome inertia, it may be necessary to reduce speed. Therefore, there must be a trade-off between speed and accuracy.

Repeatability

It is a common mistake to confuse accuracy with repeatability. Even though both are dependent on spatial resolution and mechanical inaccuracy, there is a difference. Accuracy deals with programming the robot's hand to go to a designated position. *Repeatability* expresses how close the robot's hand will actually come to the designated position while repetitively performing the task or procedure.

Good repeatability is more desirable than accuracy because inaccuracies are easier to correct. This is especially true if the inaccuracies are consistent for all movements of the robot's hand. For example, suppose that a robot is programmed to move its gripper from point A to a target point 30" (76.2 cm) away. After the robot has made the move, a measurement is taken which shows the robot actually moved 30.10" (76.454 cm). This represents an inaccuracy of 0.3 percent. If an inaccuracy of 0.3 percent is consistent for other command movements, then the programmer can compensate for this error. Adjustments for poor repeatability are more difficult. In fact, repeatability may become so poor that a more sophisticated robot is required to complete the task. However, if the error is rather small, additional tooling or alignment devices can be used to compensate.

Repeatability can change with use, especially when robots perform the same task day after day. The mechanical components are subject to wear, which increases mechanical inaccuracies. Mechanical inaccuracies, in turn, reduce the repeatability performance.

Operational Speed

Operational speed, also called *dynamic performance*, represents how fast the robot can accelerate, decelerate, and come to a stop at a given point. The two most important factors that influence operational speed are the desired accuracy and the payload. Other factors are the robot configuration and location of the tool in the work envelope.

Robot manufacturers define robot speed in different ways. It may be noted for each joint or for various groups of joints. The range of speed may be given for when there is no load, as well as for a full load.

Most people believe that robots are more productive than humans and assume that robots work at a faster pace. However, a robot's actions often are no faster than those of a human worker. In fact, a robot's movements can be many times slower than a human worker. Increased productivity results because the robot works at a constant, steady pace and doesn't stop for breaks or to eat. Therefore, basing the total cycle time for a given task on the manufacturer's speed rates may not produce the true cycle time. To arrive at cycle times that are realistic, a prototype layout may have to be built. The prototype layout should consider all the variables in the procedure (such as environment, speed of other machinery, size of the work envelope). It is important to determine how fast the end effector can move through the total cycle while still maintaining the desired accuracy and repeatability. The robot should produce a quality job in a reasonable amount of time.

Load Capacity

The *payload* is the maximum weight or mass of material a robot is capable of handling on a continuous basis. In most cases, the weight of the end effector is included in the payload. Some manufacturers' specifications indicate payload values for both an extended robot arm and for a retracted arm. Other manufacturers may list load-carrying capacity for the wrist joint and end effector, in addition to the arm. However, the two factors that greatly affect load-handling capabilities are the type of configuration and the placement of the end effector within the work envelope. The robot's load-handling capability is reduced when its boom is fully extended, compared to when the boom is retracted.

Some robots are capable of lifting as little as 1 lb. (0.453 kg), while others may have a lifting capacity of 2000 lbs. (907.18 kg). In most instances, the weight of the end effector is included in the lifting capacity. If a robot has a lifting capacity of 5 lbs. (2.268 kg) and the gripper weighs 2 lbs. (0.907 kg), the weight of a part to be moved or manipulated cannot exceed 3 lbs. (1.36 kg). Most robot manufacturers list both normal and maximum load-handling capacities.

The average weight carried by robots in U.S. industry is estimated to be 20 lbs. (9.07 kg). Approximately 50 percent of robots in industry today handle parts weighing less than 10 lbs. (4.536 kg). The average load-carrying capacity could be even less in the future because smaller robots are being constructed for assembly tasks, which rarely involve heavy payloads. For example, a large portion of the individual parts used in an automobile weigh less than 5 lbs. (2.268 kg).

4.3 Using Robots in Industry

As robot technology advances, areas where a robot might be used are limited only by our imagination and creativity. A completely automated factory is now possible and a few are in operation as a result of the advances in computing technology and sensor applications. Even now that the automated factory is a reality, engineers and technicians are needed to ensure the technology continues to function as designed. As newer robot equipment is developed and brought into the workplace, the potential uses for robots will increase dramatically.

Pick-and-Place

The process of picking up parts at one location and moving them to another is one of the most common robot applications. Placing parts or removing them from a uniform series of positions (such as palletizing or de-palletizing) is probably the most common form of pick-and-place, **Figure 4-8**. Some pick-and-place operations are used for transferring parts, **Figure 4-9**.

Figure 4-8. This robot palletizes products for storage or shipping. (FANUC Robotics)

Figure 4-9. Pick-and-place motion used to transfer products and materials. A—This robot is loading automotive wheel rims into a storage cart. (Motoman) B—This robot transfers boxes from one conveyor to another. (Pacific Robotics)

A

B

Using less-sophisticated robots for pick-and-place operations offers several advantages. They can handle fragile parts made of glass or powder metal. Those used in lightweight pick-and-place applications offer excellent speed while maintaining good accuracy and repeatability. Robots are also useful for handling heavy loads and extreme temperature items.

Machine Loading and Unloading

Another common use of robots in industry is machine loading and unloading. In production, robots load and unload parts associated with automated machining centers, such as CNC (computer numerically controlled) machines, **Figure 4-10**. A robot's machine loading and unloading capabilities can be applied to operations such as forging, injection molding, and stamping.

Industrial operations often expose workers to excessive heat, noise, dirt, and air pollution. Since these operations are considered unpleasant and are frequently dangerous, they are ideal application for robots. Replacing human operators with robots reduces the need for expensive safety equipment. Unpleasant or dangerous jobs also typically have high absenteeism rates, which contributes to low productivity and poor product quality. Substantial gains in productivity and product quality generally result when robots are properly used.

Figure 4-10. This robotic system is used for loading and unloading a CNC lathe. (Motoman)

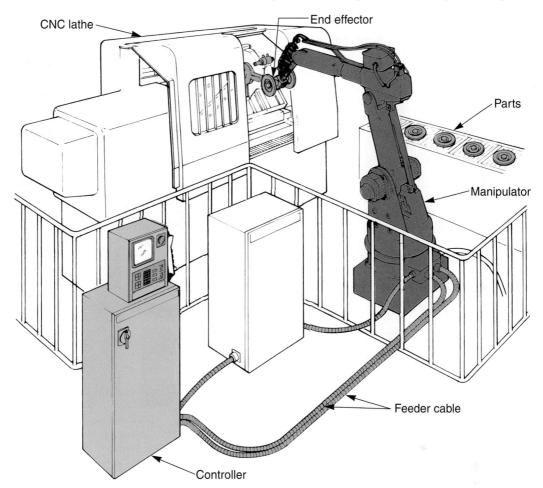

Die Casting

The first practical application of industrial robotics occurred over 30 years ago when General Motors installed the Unimate robot in a die-casting operation. The use of robots in die casting has resulted in a 200 to 300 percent increase in productivity. After robots proved useful in die casting, they were used in investment casting, forging, and welding operations.

Die casting involves pumping molten metals, such as zinc, aluminum, copper, brass, or lead, into closed dies, **Figure 4-11**. After the metal solidifies, the die is opened and the casting is removed. In this operation, the robot removes the hot casting from the die and dips it into a liquid, usually water. This process is called quenching. Next, the robot transfers the cooled part to a trimming press where excess material (flash) is removed and the part is shaped. The robot then transfers the part to a pallet or conveyor belt to be transported out of the work area.

Figure 4-11. In a die-casting operation, a robotic arm unloads the hot casting from the press and moves it to the quench tank. (Sterling Detroit Company)

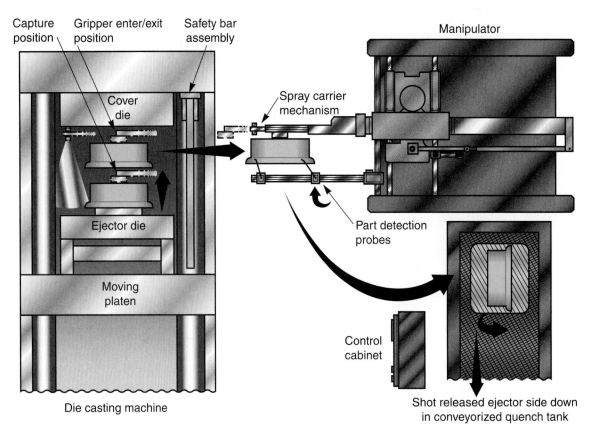

Robots can work more consistently than humans with such hot castings. The increase in productivity is the result of decreasing the total time required to produce a finished casting. The decrease in time spent in the die-casting process results in a more uniform die temperature, which reduces the amount of flash material attached to the casting. Therefore, trimming and scraping costs are also reduced.

The robots essentially perform a machine loading and unloading operation. This type of operation does not require critical or complex motions and is ideally suited for low-cost pick-and-place robots.

Welding

Robots used in resistance welding and arc welding operations comprise the third largest group of robots in industry. In resistance welding, electric current is passed between two metals, causing them to heat and fuse together at the point of the electric current, **Figure 4-12**. In arc welding, the weld is made along a joint rather than at one spot. The arc-welding process heats the metals until they melt at the joint and fuse into a single piece, **Figure 4-13**.

Figure 4-12. Robots used in industrial resistance welding applications. A—A robotic spot welder at work in an assembly operation. (FANUC Robotics) B—This robot is welding a piece of construction equipment. (Motoman)

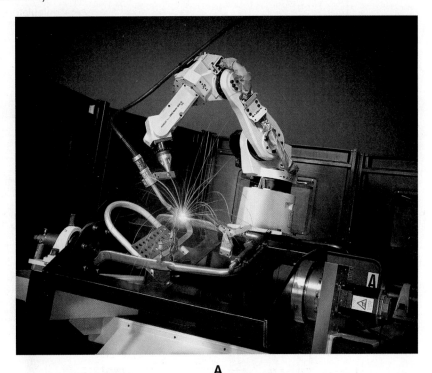

A

B

Figure 4-13. These two robots, mounted on a shuttle track for increased mobility, are being used to perform an arc welding task. (Motoman)

Robots used for spot welding (a type of resistance welding) must repeatedly move the welding gun to each specified location and position it perpendicular to the weld seam. Spot welding is used extensively in the automotive industry to weld body sections and other parts together. The workpieces must be pressed together at the spot where the weld is applied by the welding electrodes. A low-voltage direct current is passed through the parts, causing them to fuse. Robots typically used for spot welding have six or more axes of motion that allow them to approach the weld at any angle. This type of movement would be difficult and awkward for a human operator.

Many problems associated with robotic arc welding are due to poor joint alignment. Robotic arc welding requires more accurate positioning and better-fitting joints than welding done by human operators. A human operator can make adjustments if the location of joints or the gap between parts varies, but most robots cannot. To remedy the problem, robot manufacturers have developed various sensory tracking systems. One example is a robot system equipped with a visual arc sensor, **Figure 4-14**. The visual sensor tracks the joint position and compensates for deviations.

Figure 4-14. Laser seam tracking helps guide the welding laser beam along the seams. (Servo-Robot Inc.)

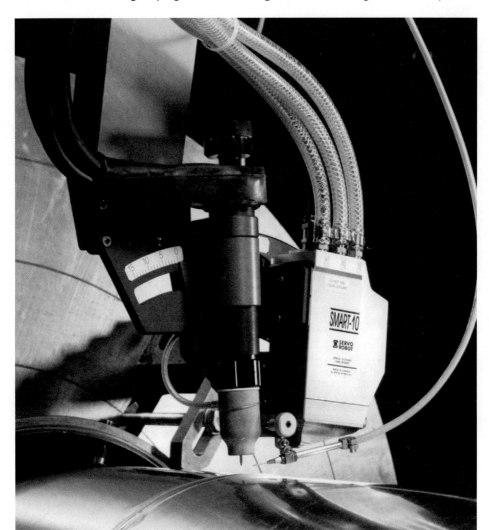

Spraying Operations

In industrial spray operations, the spray nozzle is mounted on the robot's wrist. Painting or spray finishing involves the application of a variety of paints, polyurethanes, or other protective coatings to the surface of a part or product, **Figure 4-15**. Spraying operations can also be used to apply sealant or glue to the surface of an object.

Continuous-path programming must be used for spray operations and should be performed by an experienced operator, as both intentional and unintentional movements and positions are recorded. For spraying operations, programming the robot to perform in a smooth motion is more important than the actual point locations.

Figure 4-15. These robots are spray painting a vehicle body in an enclosed work cell. (copyrighted by FANUC Robotics)

Using robots in spraying applications offers several benefits. Spray vapors are often toxic and explosive, so expensive ventilation systems must be installed for the health and safety of human workers. Since robots do not need fresh air to operate, greater concentrations of solvent can be used. With the reduction in ventilation, energy costs are lowered. However, there is the danger of an explosion caused by a spark from the robot's electrical system. In the past, most painting robots were hydraulically powered for this reason. Today, spark-free electrical robots are available for explosive environments.

The use of robots for sealing and gluing applications has increased as the quality of sealing and gluing agents and application methods have improved. The components associated with a typical robotic sealing/gluing work station are shown in **Figure 4-16**. Sealing and gluing operations require specialized end-of-arm tooling for three-dimensional surfaces, **Figure 4-17**. To ensure even distribution of the adhesive or sealing product, path control and the robot's speed must remain consistent between programmed points.

Figure 4-16. Components used in a typical sealing/gluing robotic work station. (Motoman)

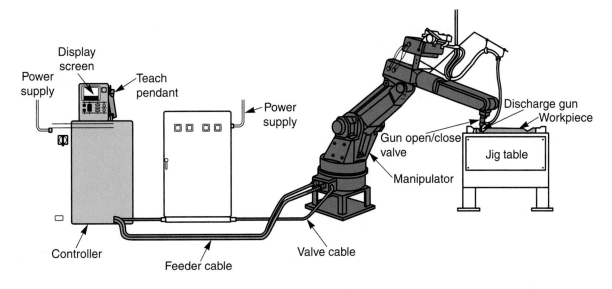

Figure 4-17. This gantry robot applies an adhesive to automotive windshields.

Machining Processes

Some of the machining processes performed by robots include routing, cutting, drilling, milling, grinding, polishing, deburring, riveting, and sanding. Robots used in these applications must have a high degree of repeatability. They must be equipped with quick-change tooling and improved sensory and adaptive control. Robots used in machining processes must also have better *fixturing* capabilities, which is the ability to provide more precise points in accurately locating parts.

Cutting applications require flexible robots with tight servo control loops, directly linked drives, and dedicated cutting axes, **Figure 4-18**. The most common types of robotic cutting applications are gas torch, plasma torch, water jet, and laser. The components used in a typical gas torch cutting robotic work station are illustrated in **Figure 4-19**. Additionally, many industrial cutting operations require special safety precautions to safeguard operators, **Figure 4-20**.

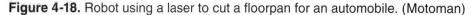

Figure 4-18. Robot using a laser to cut a floorpan for an automobile. (Motoman)

Figure 4-19. These components are associated with a typical gas torch cutting robotic work station. (Motoman)

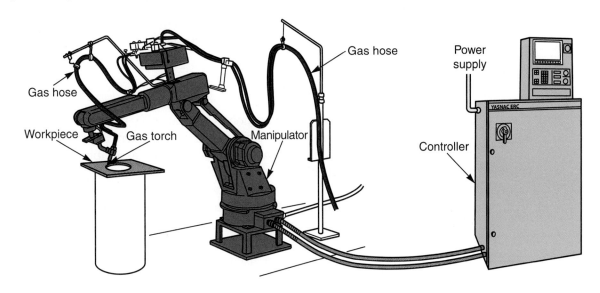

Figure 4-20. This robotic system operates within a plastic safety enclosure.

Deburring and polishing operations require robots with smooth movement and rigid wrist design. The robot control system must be capable of automatically adjusting for such things as tool wear, burr size variation, workpiece variation, and product positioning, **Figure 4-21**. Force-sensing equipment can identify contact with a workpiece and adjust tools for constant pressure, **Figure 4-22**.

Assembly

Making use of robots in some portion of the assembly process has become standard practice in industrial applications. The potential savings and increased efficiency are significant advantages, since assembly operations may account for 50 percent of the labor cost involved in manufacturing a product. Workers who perform certain kinds of assembly tasks can develop repetitive-task injuries and health problems. Psychological problems due to job stress are also associated with highly repetitive work. Such health problems make the use of automated robotic assembly systems very desirable.

Robots used for assembly work are generally small and are designed to move small parts accurately at high speeds, **Figure 4-23**. Most operations

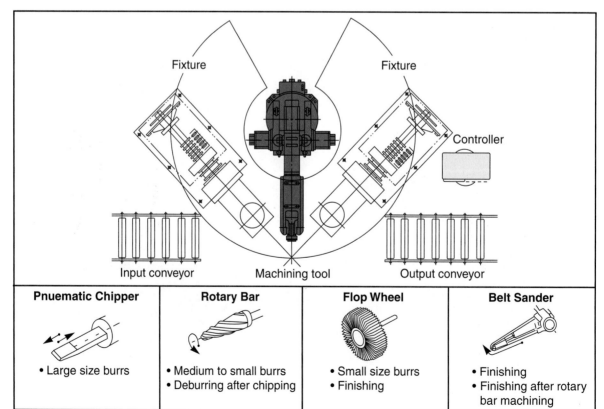

Figure 4-21. The components shown here are used for deburring and polishing. (Motoman)

Figure 4-22. Force sensors allow for corrections in tool pressure against the workpiece. (Motoman)

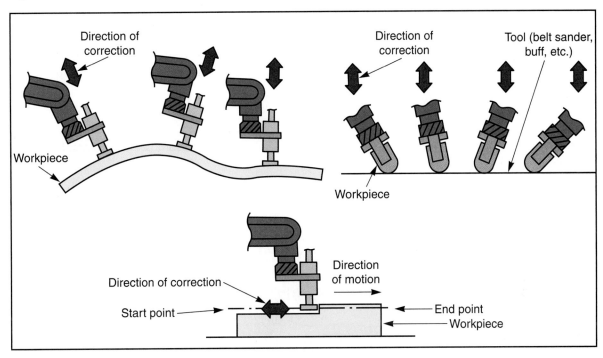

Figure 4-23. A robot inserts parts into an assembly. (Motoman)

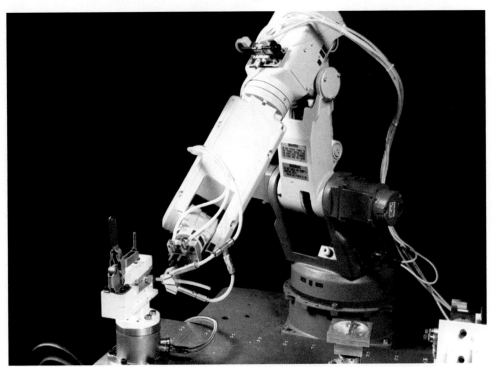

involve fitting and holding together parts and assemblies using nuts, bolts, screws, fasteners, or snap-fit joints.

Newer robots are equipped with sophisticated vision systems that can detect proper orientation of parts. Tactile sensing systems attached to the end effector are used to detect misaligned, missing, or substandard parts. The robot in **Figure 4-24** is equipped with a two-camera vision system. The

Figure 4-24. A two-camera vision system is used by this robot, which assembles air motors. (Adept Technology, Inc.)

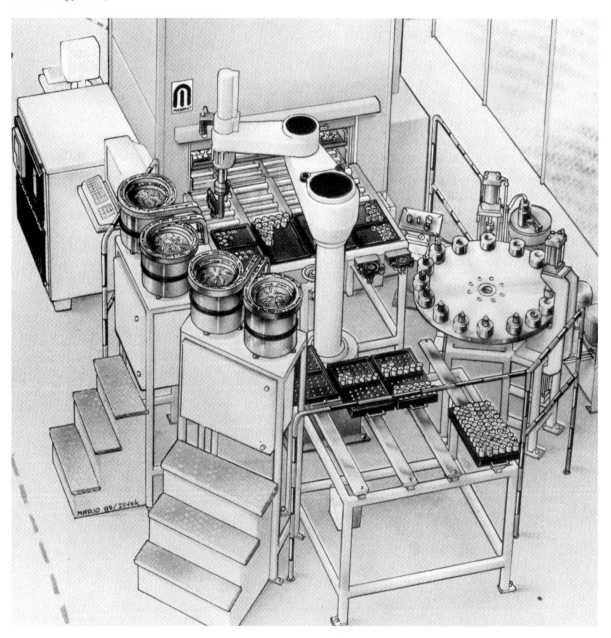

camera mounted on the manipulator guides the robot around the work cell to gather and assemble parts. The second camera is used for precise placement of a component—to within 0.0039″ (0.01 cm) of the programmed point. In this example, the motors being assembled include nearly a dozen different components. The components are fed to the robot by four vibrating bowl feeders and a paternoster (a device that holds trays of components). After assembly is complete, the robot removes the motors from the assembly area and places them on an outlet tray. This allows the operator to remove the trays without interrupting production.

To obtain the necessary production mix, the operator enters the model, quantity, and due date for each order. The robot controller uses this information to calculate an assembly schedule. It signals the paternoster to deliver the appropriate trays of components before production of a new batch begins. As a result, no changeover time is required.

Inspection

Quality assurance (ensuring products are of the highest quality) remains an essential and costly part of robotic manufacturing. The demand for non-contact gauging and inspection systems has grown steadily in recent years. At the heart of these systems is state-of-the-art electronic sensor technology, **Figure 4-25**, which checks tolerances, positioning, fixturing, and defects, among other things.

Figure 4-25. Laser vision cameras inspect welded blanks for defects, such as porosity and undercut. (Servo-Robot, Inc.)

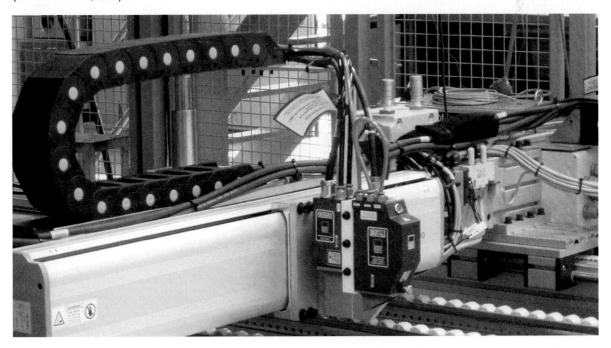

Robotic inspection systems consist of two subsystems—one accumulates data and the other analyzes the data and presents it in a meaningful way. The electronic sensing units used in these systems are set up in one of two ways. They may be mounted on the robot's hand and the robot is programmed to move the sensor along the part. The sensing unit may also be located within the work cell and inspects the parts as they move along the conveyor system.

Material Handling

Automated and robotic systems are often used for material handling. Automated control and tracking of inventory and handling parts and products increases efficiency and flexibility.

Automated guided vehicles (AGVs) are computer-controlled, battery-operated transportation devices (**Figure 4-26**). AGVs operate using one of several navigation options: buried wire guidepath, magnetic tape guidepath (inertial guidance), laser target guidance, or GPS. The type of navigation used depends on the work environment and type of tasks the AGV will perform.

Figure 4-26. This AGV transports palletized materials to a conveyor system within a manufacturing facility. (Daifuku Co., Ltd.)

Careers in Robotics: Robotics Engineer

A robotics engineer designs and maintains robotic systems, and is most often employed in the manufacturing industry. In the area of manufacturing, a robotics engineer may design new robotic systems for a production or assembly line and is likely responsible for maintaining existing robotic systems. Outside manufacturing, employment opportunities include research and robotic system development for military use, mobile exploration, medical applications, solutions for consumers, and many others.

Because robotics engineering relies heavily on computer capabilities and programming, advances in these fields create new opportunities for robotic systems and their applications. From medicine to the military, robots that were once a conceptual design are now reality due to advances in technology and the creativity of robotics engineers.

Typical education for a robotics engineer includes a graduate degree specializing in electrical or mechanical engineering and knowledge of computer science, applied science, and 3D modeling applications. To work in the manufacturing sector, education in or experience with manufacturing processes is also critical for developing and maintaining the robotics systems in a plant. A doctorate degree opens career opportunities in research and highly-specialized system development.

For safety, AGVs have bumpers and automatically stop if they contact any object in their path. AGVs are bi-directional and emit a beeping sound to warn humans of their presence. Some AGVs even signal the command center when their batteries are getting low. The computer then directs the vehicle to a battery charging station.

Service

The industrial robots described so far are limited to six or, at most, seven degrees of freedom. Their restricted movement means that the materials needed to perform a task must be brought to the robot's work envelope. *Service robots*, however, are mobile and can move to the work area to perform the necessary tasks.

One area where service robots are not only useful but also economically feasible is security services, **Figure 4-27**. Mobile robots patrol the hallways of banks, museums, government, and other high-security buildings. The command center can be located in the same building or miles away. When the operator enters the appropriate command, the robot unplugs

Figure 4-27. PatrolBot follows a programmed patrol route or schedule and provides remote video and 2-way audio to the command center. (MobileRobots, Inc.)

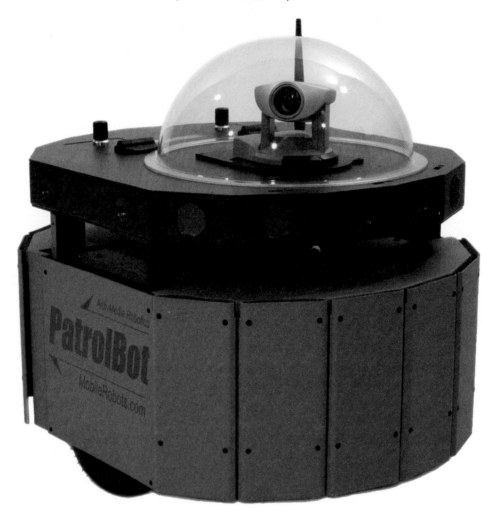

itself from its charging station and begins its rounds. As it patrols, the robot continuously samples the air and surveys the environment with its sensors. It also automatically responds to any emergency or intrusion. The robot can be programmed to stop and perform automatic surveys when suspicious conditions are detected. The information is transmitted back to the command center where the operator can see the situation on a video console.

Mobile service robots are also used in hazardous environments and assist in life-threatening situations. Mobile robots can be found in such diverse areas as the nuclear industry, security and law enforcement, the military, mining, firefighting, construction and excavation, and the removal of chemical or toxic hazardous materials.

One day, robots may perform all of the chores in your daily life, such as cleaning your home, cooking your meals, taking out the garbage, and fetching your newspaper from the front yard. In the retail market today, there are already robots you can program to clean your floors (**Figure 4-28**), mow your lawn, and monitor the security of your home (**Figure 4-29**). In his book *Robots in Service*, Joseph F. Engelberger writes that stand-alone robots able to pump gas, fill prescriptions, cook and serve foods, clean commercial buildings, and aid the handicapped and elderly are real prospects. These robots will bring the magic back to robotics. Many robotics experts agree that in the decades ahead, service will be the fastest-growing robot application.

Figure 4-28. The Roomba® can be programmed to vacuum the rooms in your home and automatically adjusts when moving from carpeting to hard floors. (iRobot Corporation)

Figure 4-29. The Spy-Cye provides a mobile video monitor in your home. You can log on to your Spy-Cye while away from home and see that everything is secure. (Educational Robot Company)

Review Questions

Write your answers on a separate sheet of paper. Do not write in this book.

1. One important application of robots is for product assembly. What are five of the key design criteria product engineers should consider when designing parts for robotic assembly?
2. Companies will often intentionally design parts for robotic assembly. Are there any advantages to this practice, even if robots are not actually installed on the production line?
3. Identify four types of safety equipment commonly used to make working around robots safer and explain how each functions.
4. Identify four primary factors that must be considered when selecting a robot to perform an industrial task.
5. Define command resolution and spatial resolution.
6. What is the spatial resolution of a robot that has a command resolution of .006″ (0.0152 cm) and a mechanical inaccuracy of .004″ (0.0101 cm)?
7. Spatial resolution may vary at different locations in the work envelope for certain robots. Which robot configuration offers the most consistent spatial resolution throughout its work envelope?
8. Briefly explain the difference between accuracy and repeatability.
9. How is the dynamic performance of a robot affected by the weight of a part handled by the manipulator?
10. What is the maximum payload of a robot with a 50 lbs. (22.68 kg) load-carrying capacity and an end effector that weighs 8 lbs. (3.628 kg)?
11. List seven major production applications for robots.
12. Can present-day robots perform every industrial task that humans now perform? Explain.
13. Which robot application has the greatest potential for growth in the future?

Learning Extensions

1. Search the U.S. Department of Labor: Occupational Safety & Health Administration Web site (www.osha.gov) for the robot safety information in the *OSHA Technical Manual*. Review the types of hazards explained in the document and the information presented on safeguarding personnel.
2. Access The Tech Museum of Innovation at www.thetech.org and visit the Robotics online exhibit. This is a very informative site that provides a wealth of information about robots and their function.

Unit II
Power Supplies and Movement Systems

Three types of motion are used in automated applications: rotary, linear, and reciprocating. These types of motion can be produced using either electrical or fluid (hydraulic or pneumatic) powered operating motors, relays, solenoids, actuators, or cylinders. The manipulator is the basic mechanical unit of a robot. It has several moving joints and performs the actual work functions of the machine.

Unit II Chapters

Chapter 5: Electromechanical Systems 115
5.1 Automated Systems and Subsystems 116
5.2 Mechanical Systems 119
5.3 Electrical Systems 119
5.4 Overload Protection 137

Chapter 6: Fluid Power Systems 139
6.1 Fluid Power System Models 140
6.2 Characteristics of Fluid Flow 144
6.3 Principles of Fluid Power 146
6.4 Fluid Power System Components 149
6.5 Hybrid Systems 167

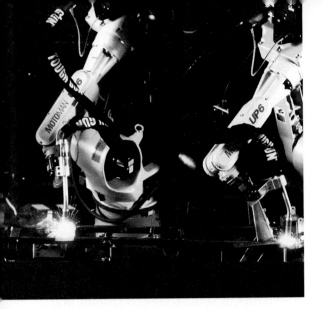

Chapter 5
Electromechanical Systems

Outline

5.1 Automated Systems and Subsystems
5.2 Mechanical Systems
5.3 Electrical Systems
5.4 Overload Protection

Objectives

Upon completion of this chapter, you will be able to:

- Discuss the use of electromechanical systems with robots.
- Explain the function of control systems used with robots.
- Summarize the characteristics of direct current, single-phase ac, and three-phase ac motors.
- Describe the type of motion that rotary electric actuators produce.

Technical Terms

alternating current (ac)
armature
bifilar construction
brushes
commutator
comparator
compound-wound dc motor
control
counter electromotive force (cemf)
cycle
cycle timing system
dc stepping motor
delay timing system
detector
detents
digital system
direct current (dc)
electric motor
electromechanical system
error detector
feedback
field winding
flux
indicator
interval timing system
light pipe
load
mechanical fuse
permanent-magnet dc motor
preloaded springs
rectification
rotor
sensing system
series-wound dc motor
servo system
servomotor
shunt-wound dc motor
single-phase ac motor
single-phase induction motor

115

slip	system	torque
squirrel cage rotor	three-phase ac motor	transmission path
stator	three-phase induction motor	universal motor
subsystem		work
synchronous motor	three-phase synchronous motor	
synchronous speed		
synthesized system	timing system	

Overview

At one time, all manufacturing operations were manually controlled. Gas-filled tubes, magnetic contacts, and electrical switchgears served as the primary control devices. Developments in solid-state electronics and miniaturization have brought a number of advances in system control. Electromechanical, optoelectronic, hydraulic, and pneumatic systems are often combined in the control of a single industrial robot. This chapter provides an overview of the types of electromechanical systems used with robots.

5.1 Automated Systems and Subsystems

Nearly all manufactured products manufactured are produced using some type of *electromechanical system*. Systems of this type transfer power from one point to another through mechanical motion that is used to do work. Punch presses that move up and down, rotating machinery, and robots are all examples of electromechanical systems.

A *system* is a combination of components that work together to form a unit. A robot is a unique type of system that may require several different types of inputs, controls, and specialized machinery for proper operation. These various components are the *subsystems* that comprise a complete robot. When subsystems are combined, the result is referred to as a *synthesized system*. Familiarity with each of the subsystems and the location of each within the overall system is important to understanding the system itself, **Figure 5-1**.

A variety of subsystems are used in virtually all automated systems. An electrical power system, for example, is needed to produce and distribute electrical energy. Hydraulic and pneumatic systems are used for motion and for system control functions. Optoelectronic systems are used for inspection operations and in many types of sensors. Mechanical systems are needed to hold objects for machining operations and to move parts and assemblies on a production line. Although each system has unique features, the basic subsystems are common to all automated systems: an energy source, transmission path, control, load, and indicators.

Figure 5-1. The basic parts of an automated system must work together or the system will not function properly.

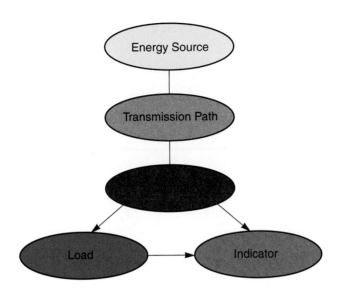

Energy Source

The energy source provides power for the system. The most common source of power for synthesized systems is *alternating current (ac)*. In alternating current power, electrons flow first in one direction and then in the opposite direction. Each repeated pattern of direction change is called a *cycle*, **Figure 5-2**. The alternating current required for an automated manufacturing system may be either three-phase or single-phase. Three-phase alternating current is ordinarily used for larger systems.

Some machines used for automated manufacturing require *direct current (dc)*. In direct current, electrons flow in only one direction. Many robotic systems use dc servo motors and controls. The process of converting alternating current to direct current is called *rectification*. Rectification is the most convenient and inexpensive method of providing dc energy to machines.

Transmission Path

The *transmission path* provides a channel for the transfer of energy. It begins at the energy source and continues through the system to the load device. This path may be a single feed line, electrical conductor, light beam, or pipe. Some systems may have a supply line and a return line between the source and the load. There may also be a number of alternate transmission paths in the system. These may be connected in series to a number of small load devices or in parallel to many independent devices.

Figure 5-2. This illustration depicts the voltage cycles in ac power. Each cycle contains two changes in direction—a high peak and a low peak.

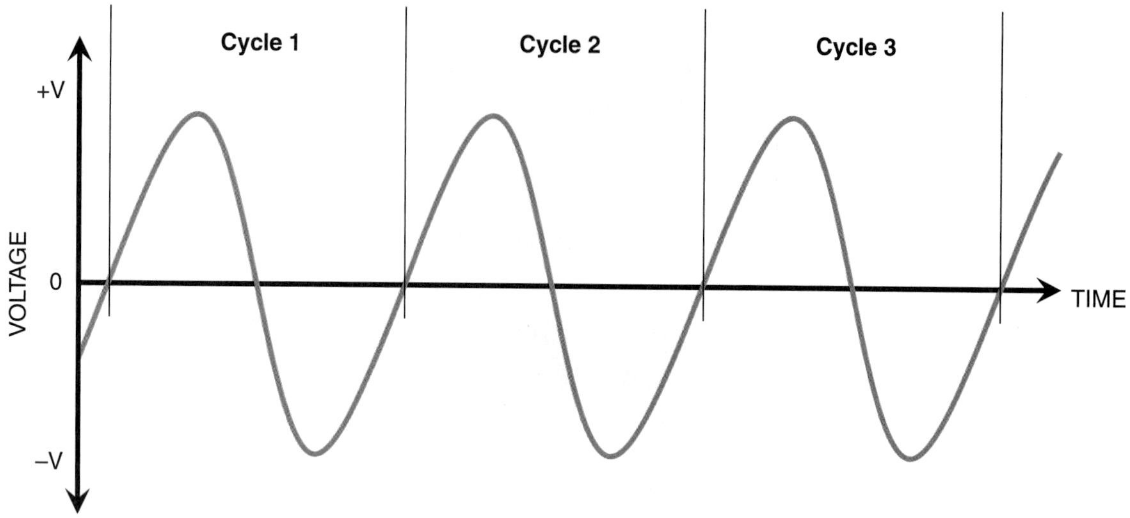

Control

The *control* alters the flow of power and causes some type of operational change in the system, such as changes in electric current, hydraulic pressure, light intensity, or air flow. Control devices operate within the transmission path. In its simplest form, control occurs when a system is turned on or off. This type of control can take place anywhere between the source and the load device.

Load

The *load* refers to a part (or parts) designed to produce work. *Work* occurs when energy is transformed into mechanical motion, heat, light, chemical action, or sound. Normally, the largest portion of energy supplied by the source is consumed by the load device. Loads include electric motors, heating systems, lighting systems, alarms, horns, and mechanical actuators.

Indicators

The *indicator* is a subsystem that displays information about operating conditions at various points throughout the system. In some systems the indicator is optional, while in others, indicator readings are essential. When operations or adjustments are critical, functionality may depend on specific indicator readings. Some types of indicators include digital meters, pressure gauges, tachometers to measure speed, and thermometers to measure temperature.

5.2 Mechanical Systems

A mechanical system produces some form of mechanical motion. Automated applications use three kinds of motion: rotary (motors), linear (relays, actuators, and cylinders), and reciprocating (certain types of motors and cylinders). The energy source for these motions can be either electrical energy or fluid power (hydraulic or pneumatic). The energy may be used to operate motors, relays, solenoids, actuators, or cylinders. As in other systems, the load is responsible for performing this action. Industrial loads are usually designed for continuous operation over long periods of time. The basic mechanical unit, such as the robot's manipulator, has several moving joints and performs the actual work function of the machine.

The transmission path of a mechanical system transfers power from the energy source to the rest of the system. The transmission path may consist of electrical conductors, belts, rotating shafts, pipes, tubes, or cables. For example, the power from an electric motor may be transferred by a belt or gear to operate a machine tool.

Control is accomplished by changing pressure, direction, force, and speed. Pressure regulators, valves, gears, pulleys, couplings, brakes, and clutches are used to control variables such as force, speed, and direction. Such changes alter the performance of the system, allowing adjustments to be made that optimize the functionality of the system.

In a mechanical system, indicators measure physical quantities. These physical quantities include pressure, flow rate, speed, direction, distance, force, torque, and electrical values. Many of these must be monitored periodically. Some indicators are used to test system conditions during maintenance. Others are designed to measure physical changes that take place. Some examples of indicators are pressure gauges, flow meters, tachometers to measure speed, anemometers to measure wind direction, and multimeters to measure electrical voltage, current, and resistance.

A sample mechanical system is a warehouse conveyor system. A conveyor moves materials, such as boxes, from one location to another. The energy source for a conveyor system is typically an electric motor. The transmission path includes pulleys, belts, or gears that connect the motor to a conveyor belt. Control is accomplished by changing speed or direction of the motor. Indicators in this type of mechanical system can measure speed or flow rate to determine the movement of boxes along the conveyor belt.

5.3 Electrical Systems

There are many different types of electrical systems used in robotics. Electrical systems include those used for sensing, timing, control, and providing rotary motion (motors).

Sensing Systems

A *sensing system* signals a response to a particular form of energy, such as light. The light source, transmission path, control, and load device are essential parts of the system. The light is produced by electrical energy and may be in the form of incandescent lamps, flames, glow lamps, electric arcs, solid-state light, or laser light. Sensing systems have become one of the fastest growing and most diversified areas of industrial robotics. New systems that combine optics, electromagnetics, and electronics have revolutionized automated manufacturing. Light systems (also called photoelectronic or optoelectronic systems) are the most common form of sensing systems. Other sensing systems include sound, pressure, flow, and magnetic sensors.

The transmission path of a sensing system is somewhat unique. In **Figure 5-3**, light energy travels in a straight line in the form of an intense beam of electromagnetic waves. If the light must go around corners or be directed to unusual locations, light pipes could be used as the transmission path. *Light pipes* are flexible, fiber-optic rods that may be extended over long distances to transfer light energy from its source to a distant location.

The *detector* of a sensing system responds to energy from the source and outputs a signal that is used to control the load device. In some sensing systems, the load may be controlled directly by light-source energy. In other systems, the light energy must first be detected and amplified. There are even applications in which the detector itself serves as the load.

Figure 5-3. This sensing system counts the units moving along the conveyor using system responses to light energy.

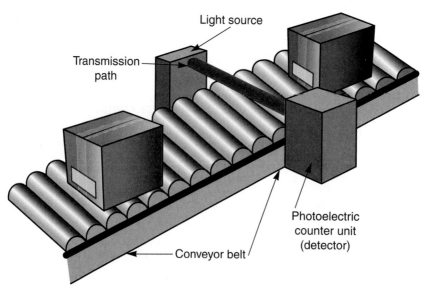

Sensing system control may be achieved by interrupting a light beam between the source and detector. Other control methods may alter the intensity, focus, shape, or wavelength of the light source. The detector's sensitivity can be adjusted to adapt it to specific operating conditions.

A common application of light-sensing systems in industrial processes is a counter (detector) on a conveyor line. An example may be a conveyor line that transports boxes from a packing area to a storage area. A light source can be placed along the conveyor line and aimed at the counter. As each box passes the light beam, the detector (counting device) adds another box to the count. Control is accomplished by breaking the light beam between the source and detector to count the number of boxes passing along the conveyor line.

Timing Systems

Timing systems turn a device on or off at a specific time or in step with an operating sequence. Types of timing systems include delay timing, interval timing, and cycle timing. *Delay timing systems* provide a lapse in time before the load device actually becomes energized. *Interval timing systems* are used after a load has been energized and operate using specified time intervals. For example, an interval timing system may be set to allow the load to remain energized only for a certain period. *Cycle timing systems* are typically more complex and may include both interval and delay timing to provide energizing action in an operational sequence.

Timing systems also include thermal devices, motor-driven mechanisms, or other mechanical, electrical, or electrochemical devices. Hydraulic, pneumatic, mechanical, heat, and electrical energy sources may be used in various combinations to power timing systems. A type of timing operation is accomplished by a computer's microprocessor. A microprocessor continually receives instructions, executes them, and continues to operate in cyclic pattern. All actions occur at a precisely defined time interval. An orderly sequence of operations, like this one, requires a type of precision timing system.

Control Systems

Control of an automated manufacturing system is carried out by human input or occurs automatically due to a physical change. During production, control systems are continually at work making adjustments that alter machine operation. The complexity of an operation determines the number of control functions needed. In many cases, several control components are used in various parts of the system.

The control unit of an industrial robot determines its flexibility and efficiency. Some robots have only mechanical stops on each axis. Others have microprocessor (computer) control with memory to store position and sequence data. Some important factors in the selection of a control unit are speed of operation, repeatability, accuracy of positioning, and the speed and ease of reprogramming.

Non-servo, or open-loop, control systems are the most basic. They use sequencers and mechanical stops to control the end point positions of the robot arm. Pick-and-place, or fixed-stop, robots use this type of control. Programmable servo-controlled robots are more complex. These continuous-path robots move from one point to another in a smooth, continuous motion.

Open-loop control systems are used almost exclusively for manual-control operations. There are two variations of the open-loop system—full control and partial control. Full control simply turns a system off or on. For example, in an electrical circuit, current flow stops when the circuit path is opened. Switches, circuit breakers, fuses, and relays are used for full control. Partial control alters system operations, rather than causing it to start or stop. Resistors, inductors, transformers, capacitors, semiconductor devices, and integrated circuits are commonly used to achieve partial control.

To achieve automatic control, interaction between the control unit and the controlled element (load) must occur. In a closed-loop system, this interaction is called *feedback*. Feedback can be activated by electrical, thermal, light, chemical, or mechanical energy. Both full and partial control can be achieved through a closed-loop system, **Figure 5-4**. Many of the automated systems used in industry today are of the closed-loop type.

In a closed-loop system with automatic correction control, **Figure 5-5**, energy goes to the control unit and proceeds to the controlled element (load). Feedback from the controlled element is directed to a *comparator*, which compares the feedback signal to a reference signal or standard. A correction signal is developed by the comparator and sent to the control unit. This signal alters the system to conform to the data from the reference signal. Systems that function in this manner maintain a specific operating level regardless of external variations or disturbances.

Figure 5-4. A typical closed-loop system, as shown in this diagram, incorporates a feedback loop.

Figure 5-5. The comparator and reference source in this closed-loop control system allow automatic adjustments to be made to a process while it is running.

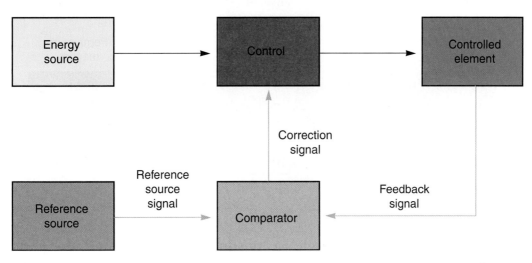

Digital Control Systems

Automatic fabrication methods, packaging, and machining operations have been improved through advances in *digital systems*. Digital (numeric) instructions are supplied by variations in pressure, temperature, or electric current. These instructions are then changed into a series of on/off electrical signals. The signals are processed by the logic circuitry of a computer and are directed to specific subsystems that perform the necessary operations.

Electrical energy energizes the load device, which performs the work. The load of a system may be electrical actuators or fluid-power cylinders designed to move parts of the robot. When appropriate signals are received, the robot performs the necessary tasks. Digital systems can control operations such as positioning and movement, clamping operations, and material flow.

Rotary Motion Systems

Electric motors convert electrical energy to mechanical motion in the form of rotary motion. Essentially, an electric motor is created by placing an electromagnet, called an *armature*, between two permanent magnets. The north and south poles of the armature are aligned with the north and south poles of the permanent magnets. When a current is passed through the armature, it becomes magnetized and begins to rotate within the magnetic field of the permanent magnets. Rotation continues until the armature's north pole is opposite the south pole of a permanent magnet, and the armature's south pole is opposite the magnet's north pole, **Figure 5-6**.

Figure 5-6. Matching magnetic poles that repel one another cause the armature to rotate in a motor.

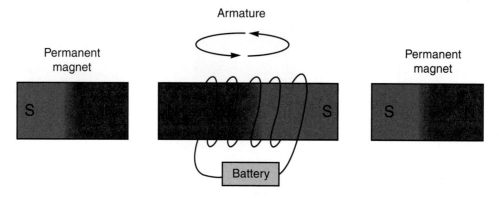

If the current through the armature is reversed, its poles will reverse and the armature will rotate again. The rotary motion, or turning force, that is produced is called *torque*. The amount of torque produced by a motor depends on the strength of the magnetic fields and the amount of current flowing through the conductors. As either the magnetic field or current increases, the amount of torque also increases.

All motors have several characteristics in common. The *stator* is the stationary portion of a motor and includes the permanent magnets, a frame, and other stationary components. The *rotor* is the rotating component of a motor and includes the armature, shaft, and associated parts. Movement of the rotor creates torque. Auxiliary equipment includes a brush-commutator assembly for dc motors and a starting circuit for single-phase ac motors.

DC Motors

DC powered motors are used when speed control is a critical factor. The basic parts of a dc motor are shown in **Figure 5-7**. A *commutator* is used to alternately switch the direction of current flow through the brushes and windings to produce a one-directional direct current (dc) through the windings of the armature. The *brushes* are usually made of carbon and allow current to flow through them. The flow of current through the brushes to the commutator and armature windings transfers current from the stationary power source to energize the rotating armature conductors.

Most dc motors use electromagnetic windings rather than permanent magnets to create the magnetic field of the stator. The coil windings wrapped around the electromagnets are called *field windings*. The interaction of the stationary electromagnetic field of the stator coils and the electromagnetic field of the rotating armature produces rotation of the dc motor.

The operational characteristics of dc motors are shown in **Figure 5-8**. As the armature rotates, it generates its own voltage called the *counter electromotive force (cemf)*. This voltage flows against the voltage coming into the motor. The amount of cemf depends on the number of rotating conductors and the speed

Figure 5-7. The basic parts of a dc motor include a brush-commutator assembly for switching the current flow.

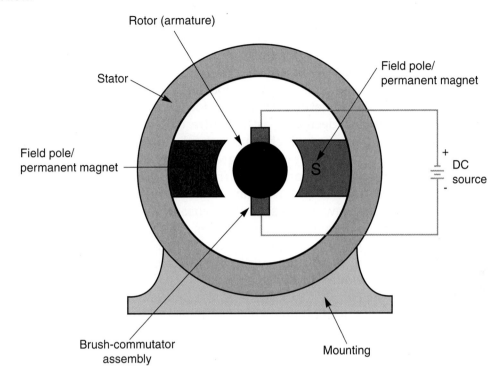

Figure 5-8. This illustration depicts the effect of mechanical load on the operating characteristics of a dc motor.

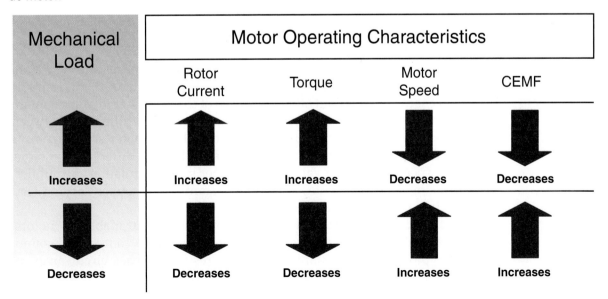

of rotation. As the mechanical load (amount of work effort demand on the machine) increases, the motor speed decreases. The opposite is also true—as the mechanical load decreases, the motor speed increases. The amount of cemf increases or decreases in direct relation to the speed of rotation.

Since the cemf flows against the supply voltage, the actual working voltage of the motor increases as the cemf decreases. When the working voltage increases, more current flows through the rotor windings. The torque of the motor is directly proportional to the amount of current flowing through the armature. Therefore, torque increases as armature current increases and decreases as armature current decreases. The amount of torque also varies with changes in load. As the load on a motor increases, torque increases to handle the greater load. The increase in torque causes the motor to draw more current from the power source.

When a motor starts, it draws a very large initial current (compared to the current draw at full speed) because of the absence of cemf. To reduce the starting current of a motor, resistors wired in series with the armature circuit are often used. Once the motor reaches full speed, these resistors are bypassed by an automatic or manual switching system. This allows the motor to produce maximum torque.

Horsepower is a measure of the amount of work performed over a specified amount of time. The horsepower rating of a motor represents the power of a motor and is based on the amount of torque produced at the rated full-load values and is a very common rating for electric motors used for robotic applications. As torque or the work requirement increases for any application, the horsepower of the motor used to drive the machine must also increase. This can be expressed mathematically as:

$$hp = \frac{2 \Pi ST}{33{,}000}$$

$$hp = \frac{ST}{5252}$$

(hp = Horsepower rating, Π = Constant, S = Speed of the motor expressed in rpm, T = Torque developed by the motor expressed in ft/lbs.)

As noted earlier, an important feature of dc motors is the ability to control speed. By changing the applied dc voltage, speed can vary from zero to the maximum rpm. Some types of dc motors have more desirable speed characteristics than others, in terms of the motor's ability to maintain a constant speed under varying load conditions. The difference between no-load motor speed (rpm) and the rated full-load motor speed (rpm) is used to determine the percent of speed regulation for different dc motors. Lower speed regulation values mean better speed regulation capabilities. Thus, a motor that operates at nearly constant speeds under varying load situations provides good speed regulation.

There are four basic categories of commercially available dc motors: permanent-magnet, series-wound, shunt-wound, and compound-wound. Each type has different characteristics that result from basic circuit arrangement and physical properties.

The *permanent-magnet dc motor* is used when a low amount of torque is needed for applications, such as motor-driven timers and printers connected to computer systems, **Figure 5-9**. The dc power supply is connected directly to the conductors of the rotor through the brush-commutator assembly. The magnetic field is produced by permanent magnets mounted in the stator.

In a *series-wound dc motor*, the armature (rotor) and field circuits are connected in a series arrangement, **Figure 5-10**. This is the only type of dc motor that also can be operated using ac power. For this reason, series-wound motors are sometimes called universal motors. In this type of motor, there is only one path for current to flow from the dc voltage source, so the field has a low resistance. Changes in load applied to the motor shaft cause changes in current though the field. If the mechanical load increases, the current also increases. This creates a stronger magnetic field.

Figure 5-9. A permanent-magnet dc motor is used for applications in which low torque is required.

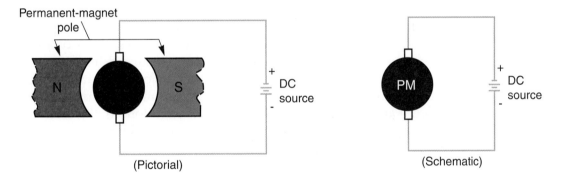

Figure 5-10. A series-wound dc motor produces high torque, but has poor speed regulation.

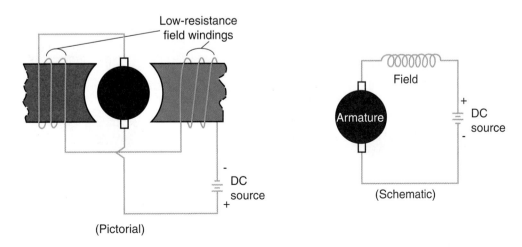

The speed of a series-wound motor ranges from very fast when there is no load, to very slow under a heavy load. Since large currents may flow through the low-resistance field, series-wound motors produce high torque. Series motors are used when heavy loads must be moved and speed regulation is not as important.

The *shunt-wound dc motor* is the most commonly used type of dc motor, **Figure 5-11**. The motor's field windings have relatively high resistance and are connected in parallel with the armature (rotor). Since the magnetic field is a high-resistance parallel path, a small amount of current flows through it.

Most of the current drawn by a shunt-wound motor flows in the armature circuit. Since the armature current has only a slight effect on the field strength, variations in load have little effect on motor speed. The field current, however, can be controlled by placing a variable resistance in series with the field windings. Since the current in the field circuit is low, a low-wattage resistor (or rheostat) can be used to vary the motor's speed. As the field resistance increases, the field current decreases. The opposite also applies—a decrease in field resistance increases the field current. Changes in the field current result in corresponding changes in the *flux* (strength) of the electromagnetic field. As field flux decreases, the armature rotates faster; as it increases, armature rotation slows.

Shunt-wound dc motors are commonly used for industrial applications because of their effective speed control characteristics. Many types of variable-speed machine tools are driven by shunt-wound dc motors.

A *compound-wound dc motor* has two sets of field windings. One set of field windings is in series with the armature, the other is in parallel with the armature, **Figure 5-12**. It has the high torque of a series-wound motor and good speed regulation like a shunt-wound motor. However, compound-wound motors are more expensive than the other types.

Figure 5-11. Shunt-wound dc motors are commonly used in industry because their speed is easily regulated.

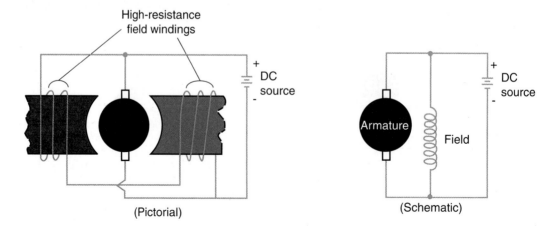

(Pictorial)   (Schematic)

Figure 5-12. The compound-wound dc motor combines the best features of the series-wound and shunt-wound motors.

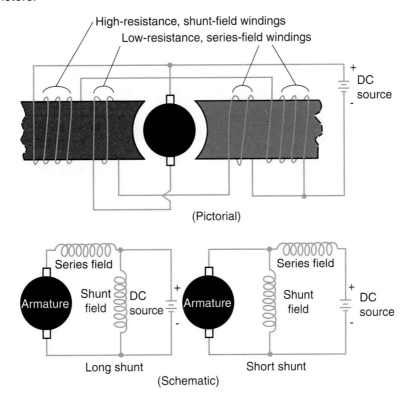

AC Motors

Single-phase ac motors are common in industrial settings, as well as in commercial and residential applications. These motors operate using a single-phase ac power source. There are three types of single-phase ac motors: universal motors, induction motors, and synchronous motors.

A *universal motor* can be powered by either an ac or dc source and is built like a series-wound dc motor, **Figure 5-13**. It has concentrated field windings and speed and torque characteristics similar to those of series-wound dc motors. Universal motors are used mainly for portable tools and small equipment.

The *single-phase induction motor* has a solid rotor, called a *squirrel cage rotor* (**Figure 5-14**). The top and bottom of large-diameter copper conductors are soldered to connecting plates. When current flows in the stator windings, a current is induced in the rotor. The stator polarity changes in step with the applied ac frequency. This develops a rotating magnetic field around the stator. The rotor becomes polarized and rotates in step with the stator's magnetic field. However, due to inertia, the rotor must be set into motion by some auxiliary starting method.

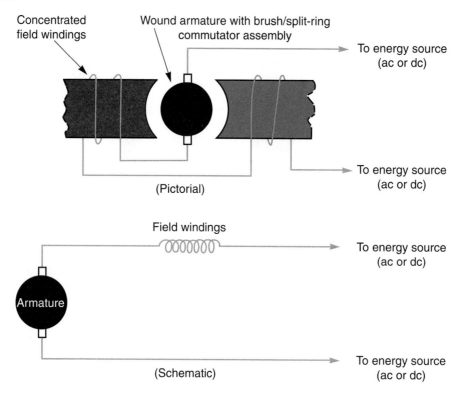

Figure 5-13. Universal motors can operate on either ac or dc energy and are often used to power portable tools.

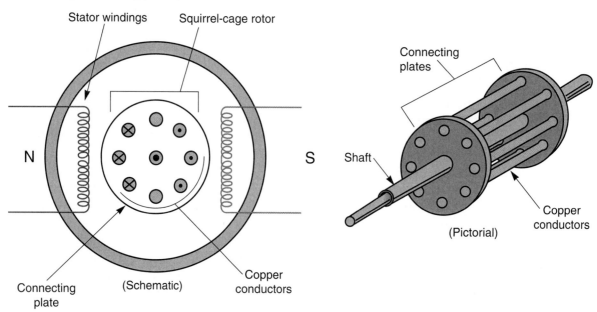

Figure 5-14. The components of a squirrel cage rotor on an induction motor include copper conductors, connecting plates, and a shaft.

The speed of an ac induction motor is based on the speed of the rotating magnetic field and the number of stator poles. The speed of the rotating magnetic field (rpm) is developed by the frequency of ac voltage (Hz) and the number of poles in stator windings. The stator speed is also referred to as *synchronous speed*. A two-pole motor operating with a 60 Hz source has a synchronous speed of 3600 rpm. For 60 Hz operation, the following synchronous speeds may be obtained:

- Two-pole, 3600 rpm
- Four-pole, 1800 rpm
- Six-pole, 1200 rpm
- Eight-pole, 900 rpm
- Ten-pole, 720 rpm
- Twelve-pole, 600 rpm

The rotor speed must be somewhat less than the synchronous speed in order to develop torque. The difference between the synchronous speed and the rotor speed is called *slip*. The greater the difference between the rotor speed and the synchronous speed, the more torque is produced. As the difference between the rotor speed and the synchronous speed becomes smaller, the percentage of slip decreases as well.

Most motors used in industry operate using a three-phase ac power source. *Three-phase ac motors* are often called the "workhorses of industry." The two basic types are induction motors and synchronous motors.

Three-phase induction motors have a squirrel cage rotor. Since three-phase voltage is applied to the stator, no external starting mechanisms are needed. Three-phase induction motors are made in a number of sizes and have good starting and running torque characteristics. Three-phase induction motors are used in industrial applications such as machine tools, pumps, elevators, hoists, and conveyors.

In **Figure 5-15**, the three-phase power source is connected directly to the windings of the stator. Phase A is connected to windings A1 and A2, which are located 180° apart. Phase B connects to windings B1 and B2, also 180° apart. Phase C follows the same pattern. The windings are distributed 360° around the entire stator. The beginnings of each phase winding (A1, B1, C1) are located 120° apart. This is equivalent to 360° divided by three (or the total number of phases). Likewise, the end of each winding (A2, B2, C2) are located 120° apart. The stator windings are energized with the three phase voltages, causing the rotor to rotate in a constant direction. Reversing any two of the winding connections (A and B, B and C, or A and C) of a three-phase AC motor causes the rotor to rotate in the opposite direction.

Three-phase synchronous motors are unique and very specialized. These motors deliver constant speed and can be used to correct power factors of three-phase systems. Direct current is applied to the wound rotor to produce an electromagnetic field. Three-phase ac power is applied to the stator, which also has windings.

Figure 5-15. A three-phase ac induction motor is widely used for industrial applications and does not require an external starting mechanism.

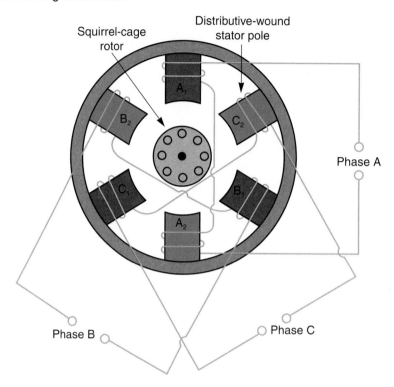

A basic three-phase synchronous motor has no starting torque. Some external means must be used to initially start it. A synchronous motor will rotate at the same speed as the revolving stator field. At synchronous speed, rotor speed equals stator speed and the motor has zero slip.

Rotary Electric Actuators

Robots require rotary electric actuators to produce rotary motion different from that produced by an electric motor. This type of rotary motion controls the angular position of a shaft. Using rotary electric actuators, rotary motion is transmitted between locations without direct mechanical linkage. Servo motors and synchronous motors are types of rotary actuators. For these applications, computer signals are applied to the actuators and translated into precise amounts of rotary motion. The electrical signals are applied to the motor causing mechanical (rotary) movement.

Servo Systems

Servo systems are machines that change the position or speed of a mechanical object. Positioning applications include numerical control machinery, process control indicating equipment, and robotic systems.

Changing the speed of a mechanical object applies to conveyor belt control units, spindle speed control in machine tool operations, and disk or magnetic tape drives for computers. In general, servo systems follow a closed-loop control path, **Figure 5-16**.

The input of a servo system is the reference source to which the load responds. By changing the input in some way, a command is applied to the *error detector*. This device receives data from both the input source and the output device. If a correction is needed, the signal is amplified and applied to the actuator. The actuator is normally a servomotor that produces controlled shaft displacements. The output device is usually a synchro system that relays information back to the error detector for position comparison.

A *servomotor* is used to achieve a precise degree of rotary motion, **Figure 5-17**. Servomotors must respond accurately to error signals and be capable of reversing direction quickly. The amount of torque developed by a servomotor must be high. There are two types of servomotors: the synchronous motor and the stepping motor.

A *synchronous motor* contains no brushes, commutators, or slip rings. It is comprised of a rotor and a stator assembly, but there is no direct contact between the rotor and stator poles. An air gap must be carefully maintained in order for synchronous motors to operate.

The speed of a synchronous motor is directly proportional to the ac frequency and the number of pairs of stator poles. Since the number of stator poles cannot be altered after the motor has been manufactured, frequency is the most significant factor in controlling speed. Speeds of 28, 72, and 200 rpm are typical; 72 rpm is commonly used in numerical control applications.

Single-phase ac synchronous motors are commonly used in low-power applications, **Figure 5-18**. This type of motor is normally limited to low-output-power applications because it develops excessive amounts of heat during starting conditions. A typical low-power application of single-phase ac synchronous motors is for precision timing circuits, where precise and constant speed is required.

The stator layout of a two-phase synchronous motor with four poles per phase is illustrated in **Figure 5-19**. This motor can start, stop, and

Figure 5-16. A typical servo system uses feedback signals to adjust the position or speed of a mechanical object.

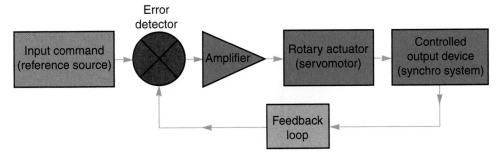

Figure 5-17. A typical servomotor is used for robot axis positioning. The controller provides signals to accurately position the axis and the motor shaft connects to the robot axis. (Adept)

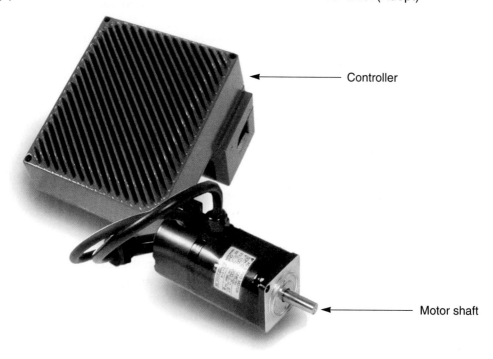

Figure 5-18. A single-phase synchronous motor has a simple construction. The rotor turns without making direct contact with the stator poles.

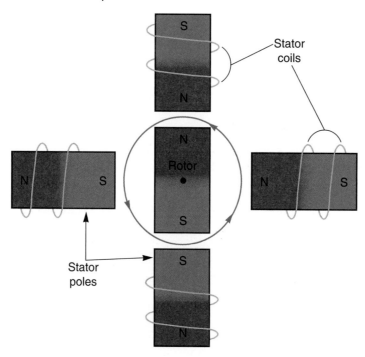

reverse quickly. In this example, there is room for 48 teeth around the inside of the stator. However, one tooth per pole must be eliminated to provide space for the windings. This leaves a total of 40 teeth. The four coils of each phase are connected in series to achieve the correct polarity.

The rotor of the synchronous motor is a permanent magnet. There are 50 teeth cast into the form of the rotor. The front section of the rotor has one polarity while the back section has the opposite polarity. The difference in the number of stator teeth (40) and rotor teeth (50) means that only two teeth of each part can be properly aligned at the same time. Since the rotor's sections have opposing polarities, the rotor can stop very quickly and can reverse direction without hesitation.

Because of its gear configuration, the synchronous motor can start in one and one-half cycles of the applied ac frequency and can stop in 5° of mechanical rotation. Synchronous motors of this type draw the same amount of current when stalled as they do when operating. This is very important in automatic machine tool applications that involve heavy mechanical loads.

Nearly all high-power servomechanisms use *dc stepping motors*, **Figure 5-20**. These motors are primarily used to change electrical pulses into rotary motion. They are more efficient and develop significantly more torque than the synchronous servomotor. The rotor is a permanent magnet.

Figure 5-19. The stator layout of a two-phase synchronous motor with four poles per phase. Poles N_1 to S_3 and N_5 to S_7 represent one phase. Poles N_2 to S_4 and N_6 to S_8 represent the second phase. (Superior Electric Co.)

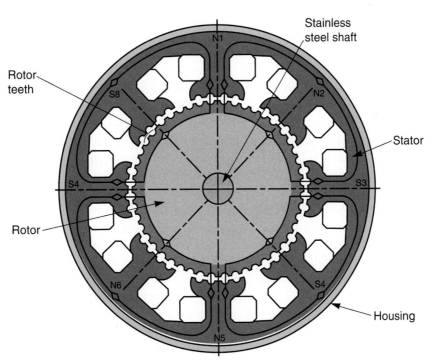

Figure 5-20. These stepping motors are used to power linear actuators. (Anaheim Automation, Inc.)

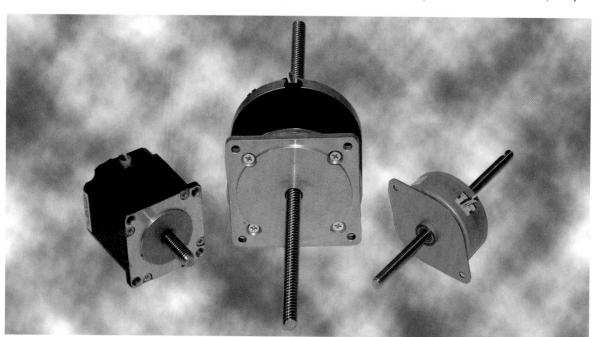

The shaft of a dc stepping motor rotates a specific number of degrees with each pulse of electrical energy. The amount of rotary movement, or angular displacement, can be repeated precisely.

The velocity, direction, and travel distance of a piece of equipment can be controlled by a dc stepping motor. Stepping motors are energized by a dc drive amplifier that is controlled by a computer system. The movement error is generally less than 5 percent per step. The construction of a dc stepping motor is very similar to the ac synchronous motor. Some manufacturers make servomotors that can be operated as either ac synchronous or dc stepping.

The stator coils are wound using *bifilar construction*, in which two separate wires are wound into the coil slots at the same time. The two wires are small, permitting twice as many turns as with a larger wire. This simplifies control circuitry and dc energy source requirements.

Operation of a stepping motor is achieved using a four-step switching sequence. Any of the four combinations of switches 1 or 2 will produce an appropriate rotor position. The switching cycle then repeats itself. Each switching combination causes the motor to move one-fourth step.

The rotor in the circuit illustrated in **Figure 5-21** permits four steps per tooth, or 200 steps per revolution. The amount of linear displacement, or step angle, is determined by the number of teeth on the rotor and the switching sequence. A stepping motor that takes 200 steps to produce one revolution, moves 360°/200 or 1.8° per step. It is not unusual for stepping motors to require eight switching combinations to achieve one step.

Figure 5-21. This dc stepping motor is similar in construction to an ac synchronous motor. (Superior Electric Co.)

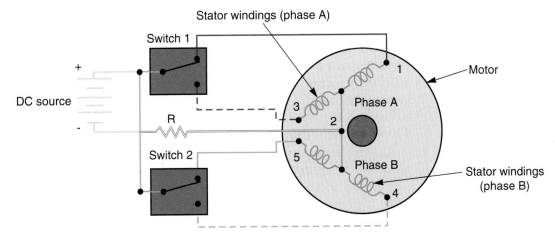

Switching Sequence*

Step	Switch #1	Switch #2
1	1	5
2	1	4
3	3	4
4	3	5
1	1	5

*To reverse direction, read chart from bottom to top.

5.4 Overload Protection

The end effectors on industrial robots must have some protection against overload. Ordinarily, a feedback signal is sent to the computer system and the manipulator is withdrawn before damage occurs. Breakaway wrist devices or rapid withdrawal of the manipulator can be accomplished using mechanical fuses, detents, and preloaded springs.

Mechanical fuses are pins or tubes that break or buckle under extreme stress. Mechanical fuses must be replaced after they perform their function, but they are not as expensive as other overload protective devices. *Detents* are two or more elements held in position by spring-loaded mechanisms. They move from their original position when placed under excessive stress. When detents move from their original position, power is removed from the machine providing overload protection. *Preloaded springs* may also be used to prevent overload conditions. Excess stress causes the spring to release and the end effector breaks away from the work area. These devices reset automatically when the overload is removed.

Review Questions

Write your answers on a separate sheet of paper. Do not write in this book.

1. What is a subsystem?
2. What is a synthesized system?
3. Name three of the subsystems utilized in the operation of manufacturing equipment.
4. Explain the function of mechanical systems in the operation of industrial robots.
5. How are sensing systems, timing systems, and control systems used in the operation of industrial robots?
6. Discuss the control system of an industrial robot.
7. Explain the difference between full control and partial control.
8. What is feedback in a closed-loop system?
9. Identify the basic parts of an electric motor.
10. Describe the relationship of load, speed, cemf, current, and torque in an electric motor.
11. What does the horsepower rating of a motor represent?
12. Identify four types of dc motors and list the characteristics of each.
13. Identify the various types of single-phase ac motors and list unique characteristics of each type.
14. What factors contribute to the speed of an ac induction motor?
15. Describe the purpose of a servo system.
16. Describe the basic construction of an ac synchronous motor.
17. What is a dc stepping motor?
18. Describe three overload protection methods used with industrial robots.

Learning Extensions

1. Select a piece of equipment or a machine and identify the parts of its system (energy source, transmission path, control, load, indicators).
2. Choose a type of electric motor and search the Internet to find the specs and operational information, such as horsepower ratings available, cost, and common applications.
3. Make a sketch of a type of electric motor that includes illustrations of a simplified rotor and stator. Describe how rotation is achieved.
4. Go to manufacturers' Web sites and obtain information on rotary actuators that may be used with robotic systems.

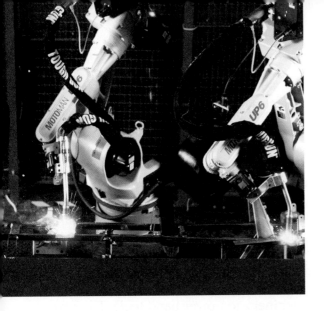

Chapter 6
Fluid Power Systems

Outline

6.1 Fluid Power System Models
6.2 Characteristics of Fluid Flow
6.3 Principles of Fluid Power
6.4 Fluid Power System Components
6.5 Hybrid Systems

Objectives

Upon completion of this chapter, you will be able to:
- Describe the characteristics of both hydraulic and pneumatic systems.
- Discuss the characteristics of fluid flow.
- Explain Pascal's Law and how it is applied in fluid power systems.
- Define the factors involved in calculating force, pressure, and area.
- Identify the components in a fluid power system and explain the use of each.

Technical Terms

centrifugal pump
conditioning
desiccant
direction control device
filter
flow control device
flow indicator
fluid motor
fluid power system
force
FRL unit
heat exchanger
inertia
lubricator
metering
non-positive displacement pump
Pascal's law
positive displacement pump
power
pressure
pressure indicator
pressure regulator valve
pressure relief valve
prime mover
reciprocating pump
resistance
rotary gear pump
rotary vane pump
strainer
turbulence
weight

Overview

Fluid power systems have a number of characteristics that distinguish them from other power systems. For example, a force as small as a few ounces can control a much larger object of several tons—this is known as multiplication of force. When operated under computer control, fluid-powered machines can move within tolerances of one ten-thousandth of an inch. Fluid power systems can also provide rotary motion at extremely high speeds or produce creeping speeds of a fraction of an inch per minute. Fluid power can be transferred to any location where a pipe, hose, or tubing can be placed, making these systems useful in transferring power to inaccessible locations over moderate distances.

Fluid power systems are used to operate a vast array of equipment—from automated manufacturing machinery to power hand tools. They are compatible with electrical, digital, or mechanical systems. Fluid power systems are efficient, dependable, easy to maintain and economical to operate for long periods of time.

This chapter provides an introduction to fluid power and discusses principles and applications of hydraulic and pneumatic fluid power systems.

6.1 Fluid Power System Models

Fluid power systems use air or liquid, or a combination of both, to transfer power. In fluid power systems, electrical energy is often used to drive a fluid pump. Electrical energy and mechanical motion are converted into the energy of a flowing fluid. Hydraulic systems use oil or another liquid, while pneumatic systems operate with air. The operating principles associated with these systems are similar in many respects. All fluid power systems consist of an energy source, transmission path, load, controls, and indicators.

Hydraulic System Model

Hydraulic systems are used for many applications in robotic system design. Hydraulic systems are relatively simple in operation, but are also heavy-duty and adaptable systems. High pressure hydraulic fluid can be transmitted through a network of hoses to operate motors, actuators, and cylinders (load devices), which then causes a robot to operate. Hydraulic fluid is controlled by control valves and other devices as it is distributed through the transmission network of hoses and tubes. Hydraulic pumps are used as the fluid source to power the system.

The energy source for a hydraulic system typically powers an electric motor, which drives a pump (**Figure 6-1**). In this model, the electric motor acts as a prime mover. A *prime mover* is a component of a power system that provides the initial power for movement in the system. The motor receives electrical energy from the source and converts it to rotary energy. The rotary energy is changed into fluid energy by the pump. A vacuum is created in the inlet port while the pump is operating, which

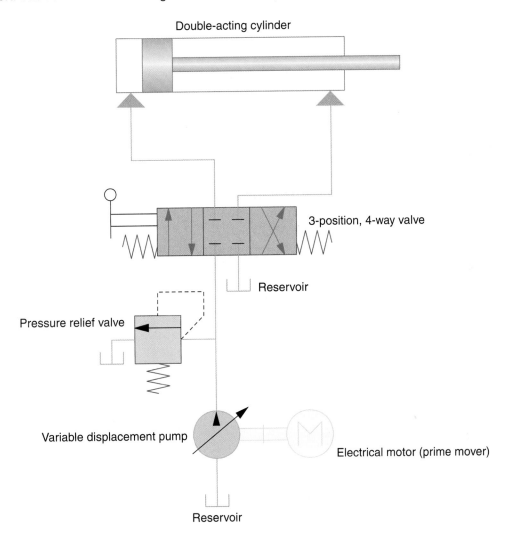

Figure 6-1. In this illustration of a basic hydraulic system, energy moves along the transmission path from the electrical motor to the pump. Fluid enters the pump at the inlet port and is forced through the outlet port. The fluid continues through the transmission path to the double-acting cylinder (load device).

draws fluid from the reservoir. The fluid moves through the pump and is forced through the outlet port and into the system.

A typical hydraulic fluid power system includes a number of control devices, **Figure 6-2**. For example, a hand shutoff valve permits control by stopping the fluid flow. A four-way, three-position directional control valve (DCV) can be positioned to allow the cylinder to extend (to restrict the flow of fluid) and lock the cylinder in place, or to shift and reverse cylinder movement. A *pressure relief valve* is a control device that protects the system from stress and damage caused by overpressure. If the hand shutoff valve is closed or the DCV is in the blocked position while the pump is running, for example, the relief valve opens and pressurized fluid returns to the reservoir.

The load device is the system component that is ultimately controlled. In the example illustrated in **Figure 6-2**, the double-acting cylinder serves as the system's load device. This cylinder changes the energy of hydraulic fluid flow into linear movement, which produces the motion of the punch-press ram.

The transmission path of a hydraulic system is often steel pipes and steel tubing, but flexible hose is also used. As the fluid passes through the transmission path, it encounters resistance to flow, which causes a buildup of system pressure. Methods in creating resistance to fluid flow in a hydraulic system include friction, control valves, actuators, and piping. Since pressure results from the resistance of fluid traveling through the system, the more resistance produced, the greater the amount of pressure. For example, as the operating speed of a pump is increased, the pressure of the system is also increased due to the greater volume of the fluid entering the system.

Pressure indicators are devices that monitor the fluid pressure within a hydraulic system. Pressure indicators or pressure gauges may be used at several points within a hydraulic system, and may be permanently affixed to certain components. Maintaining appropriate pressure levels within a hydraulic system is an important factor in consistent operation.

A major drawback of hydraulic systems is the use of petroleum-based fluids, which pose environmental dangers. Research and development efforts search for fluids that are more environmentally friendly, but still maintain the robust characteristics of petroleum-based fluids.

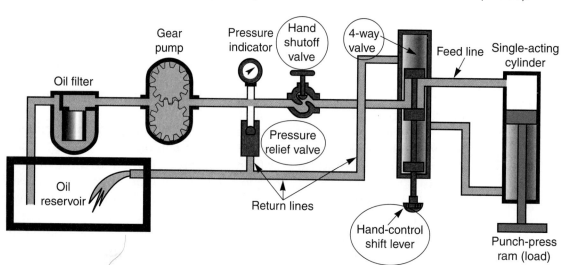

Figure 6-2. This typical hydraulic fluid power system includes several control devices (circled).

Pneumatic System Model

In a typical pneumatic system, the energy source powers a compressor which forces air into a pressurized storage tank, **Figure 6-3**. The compressor is most often driven by an electric motor or by an internal combustion engine. This storage tank holds the pressurized air and serves as the reservoir for the system. Pneumatic systems are used to power hand tools and to provide power for lifting and clamping during machining operations.

Compressed air must be conditioned to remove contaminants, such as dirt and moisture, that could damage the system components. *Conditioning* the compressed air is accomplished with an air filter that uses a condensation trap and drain. An oiler adds a fine mist of oil to the compressed air, which provides lubrication throughout the system.

Air pressure in the system must be adjusted to a specific level using the *pressure regulator valve*. Constant pressure must be maintained during system operation. Motor-driven air compressors are designed to automatically start whenever the pressure in the storage tank drops below a certain level.

The transmission path of a pneumatic system consists of piping, tubing, and flexible hoses used as feed lines. Unlike hydraulic systems, pneumatic systems do not use return lines to the storage tank. Return air is simply exhausted into the atmosphere.

A pneumatic system can have several controls throughout the system. Both hand shutoff valves and pressure relief valves control air circulating through the transmission path. Air flow can also be altered by a pressure regulator valve and a three-way valve.

The load device in a pneumatic system changes the mechanical energy of air into linear or rotary motion. An air cylinder, for example, uses linear motion to drive a punch-press ram. The load devices in some pneumatic industrial tools, such as grinders, buffers, drills, and impact wrenches, produce rotary motion.

Figure 6-3. This simple pneumatic system powers a punch-press ram actuated by a single-acting cylinder.

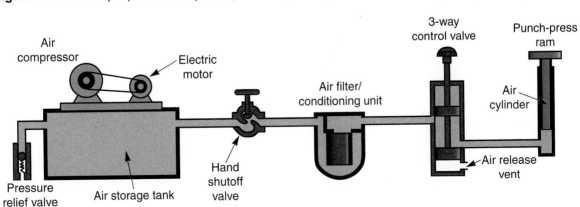

Pressure indicators are commonly added to monitor tank pressure. The output of pressure regulator valves is also monitored so that exact pressures can be determined. Test indicators are frequently used to troubleshoot faulty components.

6.2 Characteristics of Fluid Flow

In practice, fluid power systems do not achieve 100 percent power transfer from input to output. The fluid moving along cylinder walls encounters surface friction. This is one factor in the decrease of power from input to output. The power loss materializes primarily as heat. In a fluid system, friction is called *resistance*. Forcing fluid against this resistance develops system pressure. A direct relationship, therefore, exists between the amount of system pressure and surface resistance.

The relationship of resistance and pressure in a static system is illustrated in **Figure 6-4**. The pressure at Point F is zero. A pressure reading of zero could be caused by a break in the system. Other parts of the system show varying amounts of pressure in response to resistance. Point B represents the highest pressure. The full weight of the fluid occurs at this point.

Abrupt changes in the direction of flow create turbulence, which causes pressure drops. *Turbulence* refers to how the fluid moves through the fluid power system. Conditions such as size and smoothness of internal surfaces,

Figure 6-4. This illustration depicts the friction/pressure relationship in a static fluid power system.

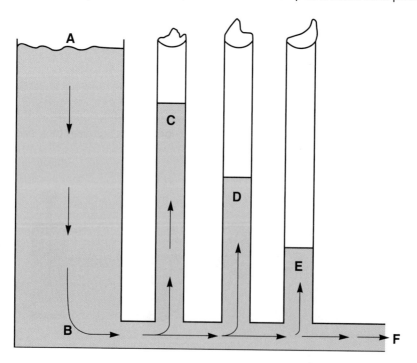

temperature of the fluid, and the location and number of valves and fittings may cause irregular flow characteristics. The corners and components illustrated in **Figure 6-5** are examples of turbulent areas in a system.

Restrictions within the system are also a source of pressure drops. Restrictions can be created by control valves, tubing length, or reduced tubing size. Smaller lines tend to increase the speed of fluid flow, which causes an increase in resistance. A pressure drop is the result of a change in the form of the fluid energy. As fluid pressure enters the system, it has the ability to perform a specific amount of work. However, as fluid passes through the system, the energy changes form. Fluid energy is lost because it is changed into heat due to friction and resistance.

In a static system, the amount of power loss due to resistance is negligible. In systems that move larger volumes of fluid over longer lines, however, the losses are more considerable. Resistance losses can generally be controlled by reducing the line length, limiting the number of bends, and using appropriately sized lines to prevent high fluid velocity. A properly designed system will help achieve and maintain a high level of efficiency.

When fluid flow ceases, the pressure reaches a stable value throughout the system. A faulty pump or loss of electricity could also cause this to occur. A break in the system line, however, normally causes a complete loss or very pronounced change of pressure.

Figure 6-5. In a flowing fluid power system, pressure drops occur either where fluid flow is restricted or where turbulence develops due to a change in the direction of flow.

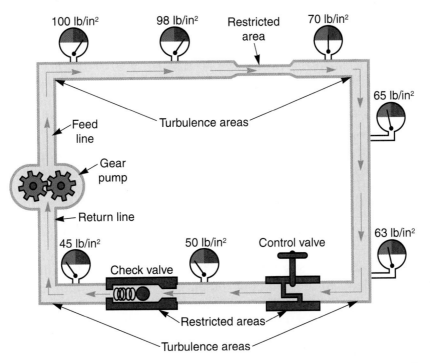

Compression of Fluids

One of the most notable differences between hydraulic and pneumatic systems is the compressibility of the fluids. All gases and liquids are compressible under certain conditions. However, hydraulic fluid (liquid) is considered virtually incompressible, whereas the air used in pneumatic systems is readily compressible. As pressure is applied to a liquid, its volume decreases by 0.5 percent per 100 psi (6895 kPa) of pressure applied. This change in volume is insignificant under normal operating conditions. Hydraulic fluids exhibit the characteristics of a solid and provide a rigid medium to transfer power through a system.

In pneumatic systems, air must first be compressed before it can be used. When air is compressed, its pressure increases, but its volume decreases. This difference in compressibility results in some differences between hydraulic and pneumatic system components. Typically, however, the basic function of a given component is similar for either system. Air is also affected by temperature—as the temperature increases, so does the volume of the air. Pneumatic systems exhibit "spongy" movement characteristics. Unlike hydraulic systems, pneumatic systems need additional components or considerations to obtain smooth operating movements.

Pneumatic systems offer a number of advantages over hydraulic systems. As the compressor produces pressurized air, it is stored in a tank and released into the system as it is needed. When the task is completed, the air does not need to be reclaimed because it is simply released into the atmosphere.

6.3 Principles of Fluid Power

In 1653, the French scientist Blaise Pascal discovered that pressure applied to a confined fluid is transmitted, undiminished, throughout the fluid. Also, this pressure acts on all surfaces of the container, at right angles to those surfaces. These findings are known as *Pascal's law* and are the basis of all fluid power systems. In the illustration shown in **Figure 6-6**, the force applied to Piston A creates pressure in the fluid that is instantly transferred to all parts of the cylinder. The pressure exerted on Piston B is equal to the pressure originating at Piston A. This is true even if both pistons are not the same physical size. The pressure acting on Piston B is also applied to the walls of the cylinder. For this reason, the walls of the cylinder must be strong enough to withstand the pressure.

Force, Pressure, Work, and Power

Force is any factor that tends to produce or modify the motion of an object. To move a body or mass, for example, an outside force must be applied to it. The amount of force needed to produce motion is based on the *inertia* (resistance to change) of the body to be moved. Force is normally expressed in units of weight.

Figure 6-6. This simple example of a hydraulic system illustrates Pascal's law. Pressure exerted by the movement of Piston A acts equally on Piston B and on all parts of the container.

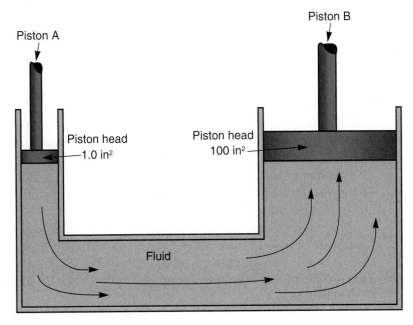

In scientific terms, *weight* is the gravitational force exerted on a body by the earth. Since the weight of a body is a force (not a mass), units of force are used to express both weight and force. The basic unit of force in the U.S. customary measurement system is the pound (lb.). In the metric system, the basic unit of force is the newton (N).

Pressure is the amount of force applied to a specific area. It is often expressed in pounds per square inch (lb./in^2 or psi) in the U.S. customary system. The pascal (Pa) is the basic unit of pressure in the metric system. Metric pressure measurements are usually expressed in *kilopascals* (kPa); one kPa equals 1000 Pa.

At sea level, the pressure of the atmosphere on the surface of the earth is 14.7 psi (101.3 kPa). In manufacturing, giant hydraulic presses can squeeze metals with a pressure as great as 100,000,000 psi. Force and pressure are simply measurements of effort. A measure of what a system actually accomplishes is called *work*. In **Figure 6-6**, work occurs when the force applied to Piston A causes it to move a certain distance. Work is expressed in foot-pounds or newton-meters (joules).

A realistic concept of work must take into account the length of time it takes to perform. *Power* is a measurement that considers the amount of work accomplished in relation to the amount of time taken to perform the work. Horsepower is used to describe mechanical power. One horsepower is equal to moving 33,000 lb. a distance of 1′ in 1 minute (or 550 lb. a distance of 1′ in 1 second) is the basic measure of horsepower. Motors are rated in horsepower.

In the simple static fluid power system illustrated in **Figure 6-7**, a 100 lb. force is applied to Piston A, which has an area of 1 in². This develops a pressure of 100 psi (690 kPa). The pressure is transferred through the fluid to Piston B. The area of Piston B is 100 in². Since pressure is transferred equally through the fluid to all parts of the cylinder, each square inch of Piston B receives 100 lb. of force. The result is 10,000 lb. (100 psi × 100 in²) of force applied to Piston B.

The distance that Piston B moves is directly proportional to the relative area of the two pistons. Moving Piston A (1 in²) 4″ into the cylinder displaces 4 in³ of fluid. The displaced volume is a product of the area of the piston and the distance it moves. In this example, 1 in² × 4″ = 4 in³ of fluid displacement. Spread over the 100 in² surface of Piston B, the fluid displacement causes Piston B to move only 1/100 of the distance traveled by Piston A (1/100 of 4″ = 0.04″ of linear travel). As a result, Piston B receives more force because of its size, but travels only a short distance.

The amount of work performed by pistons in a static system demonstrates an unusual relationship. The work done by Piston A is 400 in-lb. (100 lb. × 4″). The amount of work achieved by Piston B is also 400 in-lb. This is determined by multiplying the applied force by the distance moved (10,000 lb. × 0.04″).

Figure 6-7. In this static fluid power system, movement of the two pistons is proportional to their surface areas.

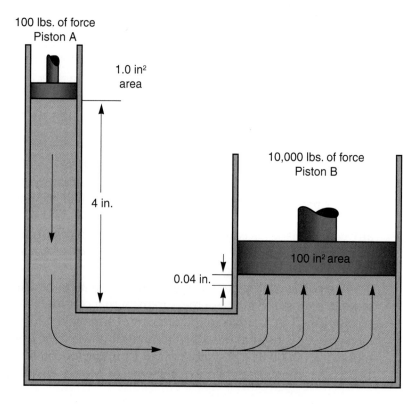

6.4 Fluid Power System Components

Hydraulic and pneumatic systems use similar, sometimes interchangeable, components. However, hydraulic components are generally larger and more rugged in construction. This is necessary because oil is more dense than air. In addition, hydraulic systems tend to be used at high pressure levels in heavy-duty automated operations. Aside from these differences, hydraulic and pneumatic equipment operate on the same basic principles and function in a similar manner.

Fluid Pumps

The pump is the heart of a fluid system. It provides an appropriate flow to develop pressure. The pump accepts fluid at an inlet port, moves it through the transmission path, and expels it from an outlet port. Gases are typically compressed into a smaller volume, which increases their pressure. Liquids flow at a faster rate to increase the pressure. The specific use of a pump within a fluid power system determines the operation it performs.

Hydraulic pumps operate continuously to keep the fluid in constant motion. A pneumatic pump, on the other hand, operates intermittently. It compresses air into a smaller volume and forces it into a tank for storage. When the tank pressure reaches a predetermined level, the pump turns off. When pressure in the tank drops to a specific level, the pump turns on again. Air compressors are often operated only for short periods.

Two general classifications of fluid pumps are positive displacement and non-positive displacement. A *positive displacement pump*, **Figure 6-8A**, has a close clearance between the moving member and the stationary components. As a result, a definite amount of fluid passes through the pump during each revolution. With a *non-positive displacement pump*, **Figure 6-8B**, the fluid is moved by the impeller blades during each rotation. The flow depends on the speed at which the blades are moving. Therefore, the amount of fluid that passes through the pump with each rotation varies.

Reciprocating Pumps

Reciprocating pumps are positive displacement pumps that use the reciprocating action of a moving piston to move fluid into and out of a chamber. A partial vacuum is created inside the chamber by the piston as it is pulled to the bottom of its stroke, **Figure 6-9A**. The intake valve opens, admitting fluid (air or liquid) into the chamber. The chamber fills to capacity by the time the piston reaches the end of its stroke. As the piston reaches the bottom, the rotary motion of the drive disk causes the piston to change direction. The discharge valve opens, the intake valve closes, and fluid is forced out of the chamber (**Figure 6-9B**).

For each revolution of the motor shaft, the piston in the reciprocating pump completes both an intake and a discharge stroke. Piston area and chamber volume are the key factors in determining the potential output

Figure 6-8. Fluid flow in positive and non-positive displacement pumps. A—Positive displacement pumps, like this gear pump, move a specific volume of fluid with each rotation or cycle. B—Non-positive displacement pumps, such as this centrifugal pump, do not move a specific volume of fluid with each rotation. Flow volume depends upon pump speed.

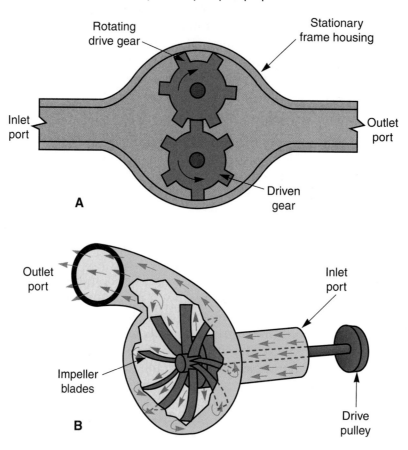

of this type of pump. In some situations, two or more stages (or cylinders) may be driven by the same motor shaft. Reciprocating pumps are often used to compress air.

Rotary Gear Pumps

Rotary gear pumps are positive displacement pumps that use rotary motion to produce pumping action, **Figure 6-10**. An external-gear rotary pump contains two gears enclosed in a precision-machined housing. Rotary motion from the power source is applied to the drive gear. As it rotates, the drive gear causes the second gear (driven gear) to turn. The teeth of the two gears mesh in the middle of the pump. The rotating gears carry fluid away from the inlet side of the pump and move it to the discharge side. The fluid is ejected through the discharge port. Because the gears mesh tightly, very little fluid returns to the inlet side of the pump.

Figure 6-9. Operation of a reciprocating pump. A—During the intake stroke, the intake valve opens and the piston moves down, drawing fluid into the cylinder. B—For the discharge stroke, the intake valve closes and the discharge valve opens. The piston moves upward to force the fluid out of the cylinder.

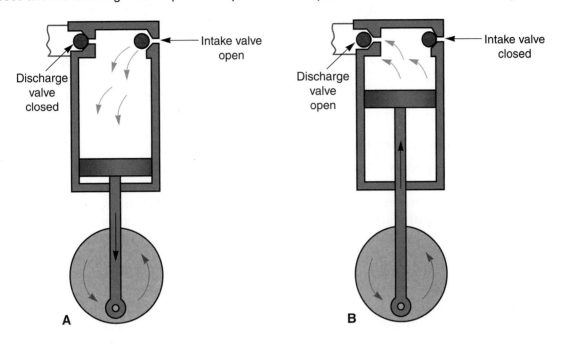

Figure 6-10. Basic construction of an external-gear rotary pump. Spaces between the teeth of the rotating gears carry fluid around the inside of the housing from the inlet port to the outlet port.

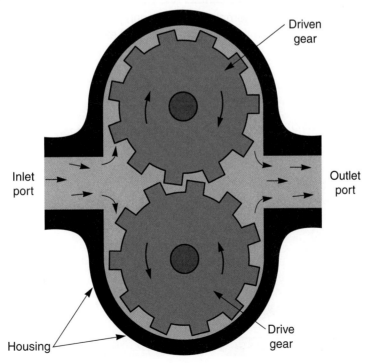

In an internal-gear rotary pump, one gear rotates within the other (**Figure 6-11**). The inner gear (idler) has fewer teeth than the outer gear (rotor). As the idler gear rotates within the rotor gear (driven gear), the gear teeth separate at the inlet port and mesh at the discharge port. Fluid is drawn through the inlet port, filling the spaces between the teeth. The fluid moves smoothly around the head crescent and is expelled at the discharge port when the teeth mesh. An internal-gear rotary pump operates equally well when rotating in either direction. Output usually ranges from 0.5 gpm to 1100 gpm (gallons per minute).

Rotary Vane Pumps

Rotary vane pumps are positive displacement pumps that use a series of sliding vanes to move fluids. The sliding vanes are placed in slots around the inside of the rotor. As the rotor turns, centrifugal force or spring action forces the vanes outward. These vanes capture the fluid as it passes by the inlet port. As the rotor continues to turn, the fluid is moved to the outlet port.

In an unbalanced (offset) vane rotary pump, **Figure 6-12**, the rotor is offset and all of the pumping action takes place on one side of the shaft and rotor. As a result, large volumes of liquid can move across the top with little or no return through the bottom. This design causes a side load on the shaft and rotor.

Figure 6-11. Fluid is moved through an internal gear rotary pump as the gears mesh and separate. Fluid is drawn in at the inlet port, flows around the head crescent, and exits through the discharge port.

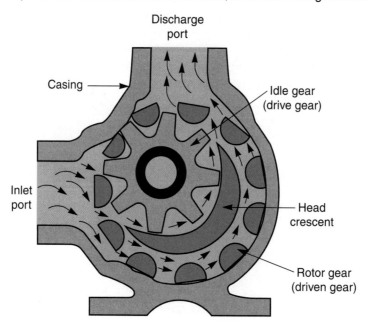

Figure 6-12. An unbalanced straight-vane rotary pump moves a large volume of fluid across the top of the chamber.

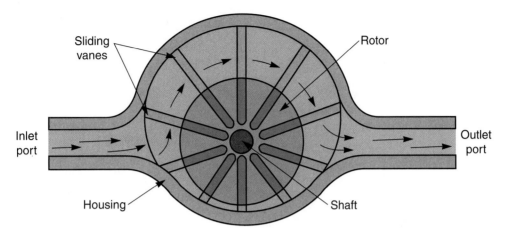

The balanced vane rotary pump is designed with the rotor in the center of its housing. Its elliptical casing forms two separate pumping areas (inlets and outlets) on opposite sides of the rotor, so that the side loads cancel out. As a result, the flow of air or liquid is smoother than with an unbalanced type of pump.

Centrifugal Pumps

The *centrifugal pump* is a non-positive displacement pump that moves a varying amount of fluid with each rotation using an impeller blade. The clearance between the impeller blade and the housing allows fluid from the inlet port to remain in the pump, even though the impeller blade is rotating. Therefore, the amount of fluid that leaves through the outlet port is not directly related to the amount of fluid that enters through the input port.

Volume depends on the rotational speed of the pump and the amount of resistance in the feed line connected to the outlet port. Increased resistance may cause fluid flow to slow or even come to a complete stop. In this case, any fluid in the pump simply rotates inside without being expelled. When this occurs, the operating efficiency of the pump drops to zero. Increasing the pump speed can solve this problem. As a general rule, centrifugal pumps are only used for transferring large amounts of fluid at low pressure.

A volute pump is a centrifugal pump with a spiral-shaped housing, **Figure 6-13A**. When fluid enters the inlet port, the revolving blades of the impeller cause rotational motion. Centrifugal force moves the fluid out toward the wall of the housing. The fluid circulates in a spiral path to the outlet port. The inlet port continually replaces the fluid expelled from the outlet port to maintain continuous fluid flow.

In an axial-flow pump, **Figure 6-13B**, the blades of the impeller maintain fluid flow along the rotational axis of the drive shaft. The axial-flow pump is

Figure 6-13. Two types of centrifugal pumps. A—A volute pump has a spiral housing that directs the flow of fluid to the outlet port. B—An axial-flow pump has an impeller that moves fluid along the axis of the drive shaft.

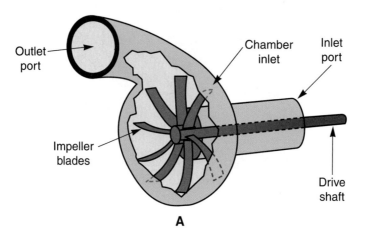

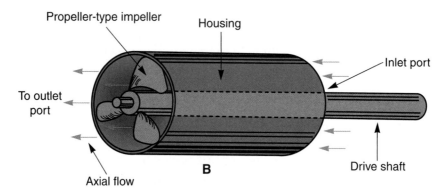

also referred to as a propeller pump because these pumps operate in a very similar manner. The axial-flow pump is versatile because the flow and pressure amounts can be altered by changing the pitch of the propeller to meet various conditions. Common applications for axial-flow pumps include irrigation, drainage, fluid mixing or circulation, and sump pumps.

Fluid Conditioning Devices

Both hydraulic fluid and air must be conditioned before being processed through a fluid power system. Conditioning devices prolong the life of fluid power systems by removing foreign particles and moisture. The number of conditioning devices used depends on the system. A simple hydraulic system may use only a line filter or strainer. More sophisticated systems may require filters, strainers, and heat exchangers. Conditioning in pneumatic systems is more complex. Devices in these systems filter the air to remove both dirt and water, regulate the pressure level, and add oil as a lubricant.

Hydraulic Conditioning

The number of components, types of control devices, and operating environment are major considerations in hydraulic fluid conditioning. For systems with manually operated control valves used in a clean environment, a simple intake strainer may be enough. However, systems with precision control valves that operate for long hours in a dirty environment may require micron filters and several strainers.

Strainers are in-line devices that capture large particles of foreign matter. They contain a stainless steel screen with 60 to 200 wires per square inch. Strainers are frequently placed in the filler opening of a reservoir, an air breather, and a pump inlet feed line.

Filters remove very small pieces of debris and provide a finer grade of fluid conditioning. Filters are typically made of some porous medium, such as paper, felt, or very fine wire mesh. Filter ratings range from 1 to 40 microns. The micron rating refers to the size of particles permitted to pass through the filter.

In-line and T filters are commonly used in hydraulic systems. T filters have a removable bowl or shell that contains the filtering element. A bypass relief valve opens when the filtering element becomes clogged and restricts flow. In-line filters must be removed when the element is cleaned or replaced. Bypass relief valves can be used with this type of filter, depending on its application.

Some hydraulic systems use *heat exchangers* to maintain a constant fluid temperature. Hydraulic machinery that operates near a furnace or is used near hot metal often requires heat exchangers to cool the fluid. Heat exchanger devices include forced-air fan units, water-jacket coolers, or gaseous cooling units. Since a hydraulic system produces heat during normal operation, cooling is needed more often than heating. Heating the fluid is required only for portable systems during cold starts.

Pneumatic Conditioning

To condition fluid in pneumatic systems, several different types of devices are used. Filtering must remove moisture, as well as foreign particles. In-line and T filters contain chemical elements made of a *desiccant*, which is a very dry substance designed to attract moisture. Desiccant elements often require periodic recharging, a heating process that dries the element. Some T filters contain a glass-bowl moisture trap at the bottom of the filter. A drain valve allows accumulated moisture to drain from the trap.

Pressure regulation is also necessary in pneumatic systems. After air passes through a T filter, it moves to a regulator valve. The movement of air through this valve can be controlled with an adjustment screw. The line pressure from the storage tank can be set to the desired operating level.

The pressure regulator valve creates a balance between atmospheric pressure and system line pressure. Atmospheric pressure reaches the top of the diaphragm through the vent. System pressure is applied to the bottom of the diaphragm. Turning the adjustment screw adds mechanical pressure to the atmospheric pressure. When this combined pressure exceeds the system pressure, the diaphragm is pushed down. This action opens the poppet valve and allows air from the storage tank to enter the system lines, **Figure 6-14A**. When the system pressure becomes greater than the combined mechanical and atmospheric pressure, the diaphragm is pushed up. This action closes the poppet valve to maintain pressure at the set level, **Figure 6-14B**. The pressure adjustment screw can be used to alter the pressure at which the valve opens. Regulators may be used in several places within a system.

Lubricators are conditioning devices that add a small quantity of oil to the air after it leaves the regulator, **Figure 6-15**. This lubrication helps the valves and cylinders in pneumatic systems operate more efficiently and last longer. In many pneumatic systems, an *FRL unit* combines the air filter, regulator, and lubricator components, **Figure 6-16**.

When air enters at the inlet port, it flows into the narrowed area, called the venturi, causing the air flow velocity to increase and pressure to decrease. Pressure in the venturi is lower than in larger areas of the pump. As a result, oil is forced from the glass bowl up the oil tube and moves to the top of the unit. The needle valve can be adjusted to regulate the oil flow so that small droplets fall into the throat. Air velocity at the bottom of the throat breaks the droplets into a fine mist that mixes with the air. Finally, the lubricated air passes through the outlet port.

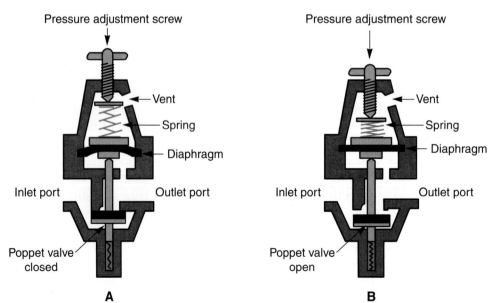

Figure 6-14. Operation of a pneumatic system pressure regulator valve. A—When system pressure exceeds atmospheric pressure, the diaphragm is pushed up, allowing the spring-loaded poppet valve to close. B—If system pressure drops below atmospheric pressure, the diaphragm is pushed down, opening the poppet valve to admit more air from the receiving tank.

Figure 6-15. A pneumatic lubricator atomizes drops of oil to create a fine mist that mixes with the compressed air supply.

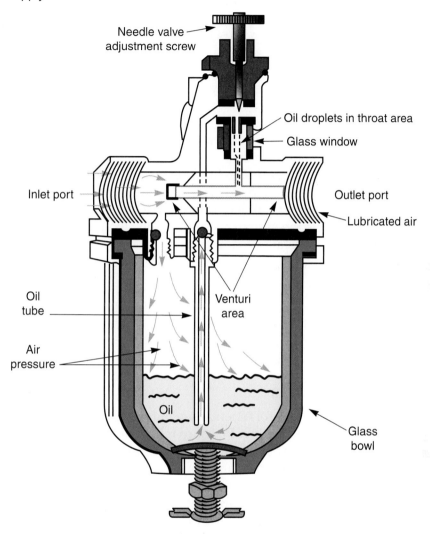

Figure 6-16. This pneumatic fluid power system has an FRL unit that combines the filtering, regulating, and lubricating devices.

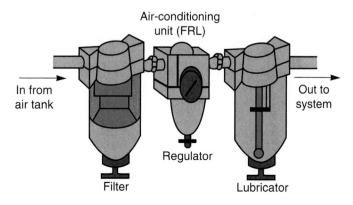

Transmission Lines

Transmission lines can be made of rigid metal piping, flexible metal tubing, or flexible hose. Rigid lines are used for permanent installations where no vibration occurs and are capable of withstanding a good deal of abuse. Rigid lines are economical, but do require additional fittings that add weight to a system.

Steel tubing is extensively used as the transmission line in hydraulic systems. It is lighter than rigid steel piping and can be bent into various shapes, which reduces the number of fittings required. Steel tubing is also more tolerant of vibration, which makes it well suited for automotive and aerospace applications.

Flexible transmission lines are made in a variety of types and sizes. The type of line used depends on the system type and its function. The type of tube or the inner lining (reinforcement material) and outside cover material used affects the system pressure limits, temperature operating range, and the hose's resistance to exposure.

Control Devices

Control is achieved by devices that alter the pressure, direction, and volume of fluid flow. Control devices are used in several different places within a system; the actual location is determined by the specific function of each control device.

Pressure Control

Pressure control functions include relief, reduction, bypass, sequencing, and counterbalancing. Pressure relief valves in hydraulic systems unload the output of a positive displacement pump back into the reservoir when the pressure rises to a dangerous level. In this case, the pressure relief valve serves as a safety device.

In pneumatic systems, pressure relief valves are used to control small amounts of air. Excess air in the system is released into the atmosphere. The output port of a relief valve may be altered in size to maintain the pressure requirements of the system.

Pressure control can also be used to establish operating sequences. Pressure relief valves can direct pressure at specified levels in a predetermined sequence. When the main system pressure overcomes the valve setting, pressure shifts to a different port. A sequencing valve can be used to force two actuators to be operated in sequence. When fluid flow to the first cylinder reaches a preset pressure, the valve allows fluid flow into a branch circuit that extends the second cylinder.

Direction Control

Direction control devices are used to start, stop, or reverse fluid flow without causing a significant change in pressure or flow rate. One-, two-, three-, and four-way valves are commonly used. These valves may be actuated by pressure, mechanical energy, electricity, or manual operation. The control action of a directional valve occurs in different ways.

One-way valves, also called check valves, operate on the seated-ball principle—they permit flow in only one direction. Pressure applied at the inlet port drives the ball away from its seat. This opens the flow path and allows fluid to pass through the valve, **Figure 6-17A**. Pressure applied at the outlet port forces the ball into its seat and prevents flow, **Figure 6-17B**. Valves of this type are often used to permit free flow around controls when the flow direction is reversed.

Two-way valves are installed in transmission lines to permit flow or to shut it off. These valves use gates, plugs, discs, spools, or other precision-machined objects to either enable or block fluid flow. For example, the ball in a ball-type control valve can be rotated manually using an outside handle. Flow is enabled when the handle is positioned parallel with the transmission line. Turning the handle 90° stops flow. This type of valve is used primarily for high-pressure applications.

Three-way valves are used to allow shifting between two different sources of pressure or to direct pressure to alternate devices. These valves are often used to alter cylinder operation or to control hydraulic or pneumatic motors. Valves of this type can be actuated mechanically, manually, or electrically to change hydraulic pressure. Basic designs include shifting spool valves, poppet valves, sliding-plate shear-seal valves, and rotary-plate shear-seal valves (**Figure 6-18**).

Figure 6-17. A check valve permits flow in only one direction. A—Fluid enters the inlet port and pushes the spring-loaded check ball to allow free-flow through the valve. B—Increased pressure from the outlet side pushes the check ball into the closed position, preventing backward flow through the valve.

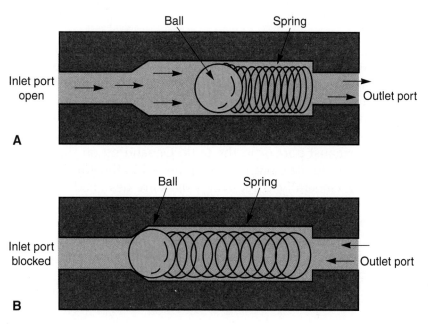

Figure 6-18. Basic types of three-way valves used for fluid control. A—Shifting spool valve. B—Poppet valve. C—Sliding-plate shear-seal valve. D—Rotary-plate shear-seal valve.

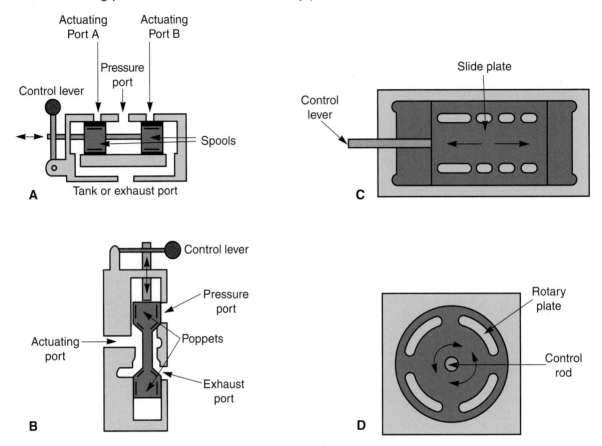

Figure 6-19 shows the basic operation of a three-way shifting spool valve. When the manual control shaft is pushed to the right, the spools also shift to the right. Fluid flows through the pressure port and the actuating port, applying pressure to the actuating device controlled by the valve. When the shaft is pulled to the left, the spools shift to the left. This action cuts pressure flow to the actuator and releases it either through the exhaust port or to the tank. Depending on the situation, pressure is directed to the storage tank or vented to the atmosphere through a relief valve. Generally, three-way valves are designed for two-position operation. Some three-way valves have a third position (neutral or off), which increases control capabilities.

Four-way valves are used to start, stop, or reverse the direction of flow, **Figure 6-20**. They are used to control forward and reverse actuation of a double-acting cylinder or to reverse the rotation of a fluid motor. The valve can be regulated mechanically, manually, or electrically.

Figure 6-19. A spool-type three-way valve. A—When the spools are moved to the right, pressure flows to the actuating port. B—When the spools move to the left, pressure is directed to an air tank or exhausted to the atmosphere.

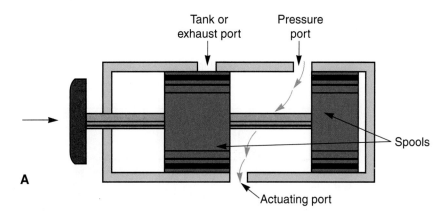

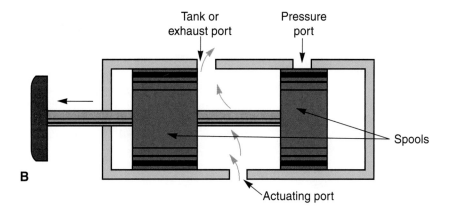

Flow Control

Flow control devices alter the volume or flow rate of the fluid. The rate at which fluid is delivered to the load of a system determines its operational speed. The speed of an air motor, for example, is dependent on the flow rate of the fluid. By altering the flow rate, the speed of the motor can be controlled.

Cylinder actuating speed is also controlled by the rate of fluid flow. To alter the linear motion of a cylinder, fluid may be controlled at the input feed line or at the return line. Controlling the rate of flow is also known as *metering*.

The operation of a flow control valve is illustrated in **Figure 6-21**. Fluid flow is controlled from the pressure-flow connection to the free-flow connection. The flow level from pressure flow to free flow is adjusted by a needle valve. Flow from the free-flow connection to the pressure-flow connection forces the check valve ball to move away from its seat, which allows free (uncontrolled) fluid flow.

Figure 6-20. In its simplest form, a four-way valve has five working connections: a pressure inlet port, two actuator ports, and two exhaust ports. Position 1—The feed line to the Pressure Port is off. Position 2—Note the direction of flow from the pressure port to Actuating Port A. Exhaust from Actuating Port B is released through Exhaust Port 2. This moves the piston in a double-acting cylinder in one direction. Position 3—Note the flow from the Pressure Port to Actuating Port B. Exhaust from Actuating Port A is released through Exhaust Port 1. This reverses the flow to the cylinder, causing the piston to move in the opposite direction.

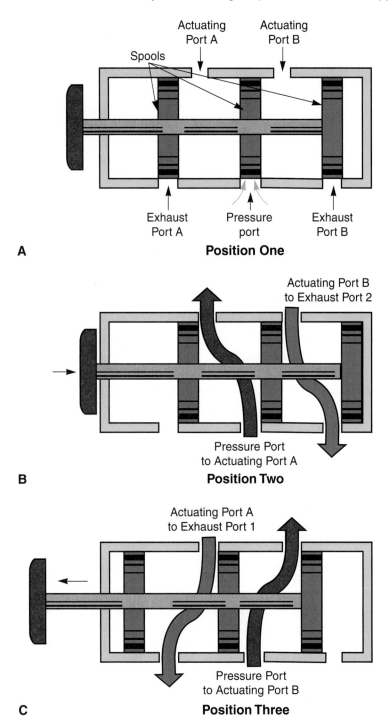

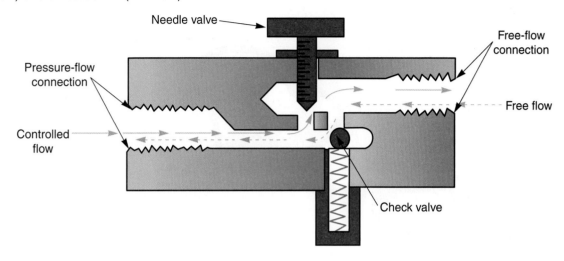

Figure 6-21. In this flow control valve example, fluid flow is adjusted either by the needle (controlled flow) or the seated ball (free flow).

Load Devices

The term *actuator* is often used to identify the load device. Fluid power systems produce work in the form of mechanical motion, provided by either linear or rotary motion. Similar operating principles apply to both hydraulic and pneumatic actuators.

Linear Actuators

Cylinders used as linear actuators develop the force needed to lift, compress, hold, or position objects (**Figure 6-22**). To produce linear motion, hydraulic fluid or air is forced into a cylinder under pressure. A piston inside the cylinder moves as pressure is applied to it. The amount of force created depends on the total area of the piston (in² or cm²) and the amount of fluid pressure (psi or kPa) applied.

Single-acting cylinders have only one input port. The piston moves when fluid is forced into the input port under pressure. A weight (the load) is lifted by the force exerted on the piston by the fluid, **Figure 6-23A**. The combined load includes the weight being lifted, the friction between the piston and cylinder walls, and the heat developed by that friction. To return the piston to its original position, the pressurized flow of fluid is stopped by closing a valve. The fluid under the piston is released by opening the valve's exhaust port. The weight of the load causes the piston to retract, **Figure 6-23B**.

Double-acting cylinders can move in two directions. Two ports are needed because fluid is supplied to one port and is simultaneously expelled from the other. Retracting the piston is accomplished by reversing the fluid flow. Applications for double-acting cylinders include punch presses, rolling mills, machine-tool clamps, paper cutters, and robot actuators.

Figure 6-22. In this pneumatic system, the cylinder is the load device and produces linear motion to shift packages from one conveyor to another.

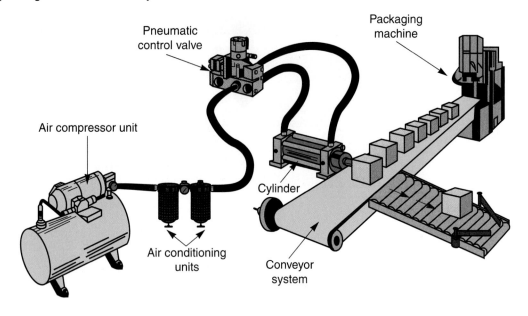

Figure 6-23. Operation of a single-acting cylinder. A—Fluid flowing under pressure through the actuating port forces the piston and its load up. B—Weight of the piston and load causes the piston to retract when pressure is released.

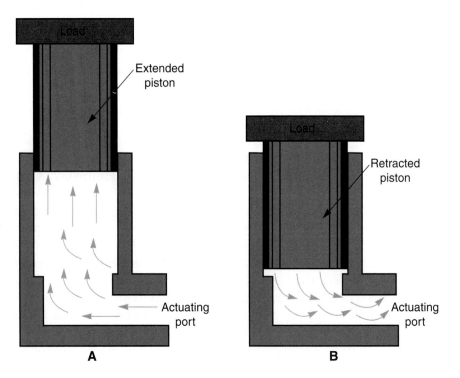

In **Figure 6-24**, fluid is applied to the right side of the piston and removed from the left side. This forces the piston to move to the left. Switching the fluid flow causes the piston to move to the right. The double-acting cylinder in the illustration is a differential type. The retracting force is somewhat less than the extending force. This is because the area of the piston on the retraction side is reduced by connection of the piston rod. A non-differential cylinder has rods extending from both ends of the piston. Cylinders of this type can provide an equal force in either direction.

Rotary Actuators

Rotary actuators produce a limited amount of rotary motion (twisting or turning) in either direction. The rotary actuator shown in **Figure 6-25** applies fluid to Port A, which causes the rotor to move in a clockwise direction. Counter-clockwise rotation occurs when fluid enters Port B and is expelled from Port A. A single-vane rotary actuator can turn in either direction through an arc of approximately 280°. A double-vane rotary actuator has twice the turning power, but can only turn approximately 100° in either direction. Actuators of this type are used for lifting or lowering, opening or closing, and indexing operations. One example of this application is the reciprocating operation of a punch press.

Figure 6-24. Operation of a double-acting cylinder. A—Fluid under pressure from the pump extends the piston. Fluid behind the piston is displaced and flows to the reservoir. B—Reversing fluid flow through the ports retracts the piston.

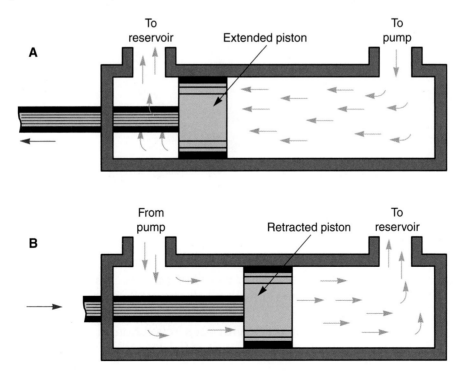

Figure 6-25. Single- and double-vane rotary actuators. A—Single-vane rotary actuators can rotate through a larger arc. B—Double-vane rotary actuators have twice the turning power, but permit less movement in either direction.

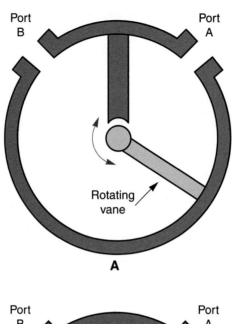

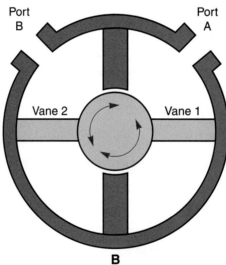

Fluid Motors

Fluid motors convert the force of a moving fluid into rotary motion through the use of vanes, gears, or pistons. Fluid motors are classified according to the type of fluid displacement they use. Gear, vane, and piston motors usually have a fixed displacement. They accept and move only a certain amount of fluid with each revolution or cycle. Operating speed depends entirely on the amount of fluid supplied by the source. In variable displacement motors, the amount of fluid circulated can be changed. The piston motor is a variable displacement motor. The length of its stroke is altered to produce variable displacement. Its speed can be changed by an external adjustment. Operating speeds of up to 3000 rpm are typical.

The turning capability of a fluid motor is a measure of torque. The torque is equal to the developed force multiplied by the radius of the rotating arm. The output power of a fluid motor is expressed as a horsepower rating.

Gear pumps can be used interchangeably as fluid motors. They are capable of operating at speeds up to 5000 rpm. Both internal and external gear pumps are currently available.

Indicators

The most essential measurement in a fluid system is pressure. Pressure indicators play an important role in overall system performance. Regulators and pneumatic receiver tanks often use permanent pressure indicators or gauges. A wide range of pressures must be measured. Negative pressures (vacuums) as low as 0.00002 psi (0.14 Pa) and positive pressures as high as 1,000,000 psi (6895 MPa) must be measured. Reading such a wide range of pressures requires a number of different devices.

Fluid systems also require measurements of flow and temperature. The readings from these indicators are valuable in analyzing system efficiency. Flow and temperature readings are only referenced periodically, so these indicators are not usually permanently installed.

In pressure indicators, an element physically changes shape to show pressure changes. For example, spiral and helix coil elements uncoil when pressure is applied. The Bourdon tube element tends to straighten. The physical change causes movement of an indicator on a scale or a stylus on a paper chart.

Flow indicators are primarily used to test flow rates from pumps and at the inlet and outlet ports of actuators. By monitoring flow rates, system efficiency is measured and maintenance problems are reduced.

6.5 Hybrid Systems

A number of industrial systems produce mechanical energy by combining fluid power and electrical power systems. Hybrid systems of this type play an important role in automated manufacturing, including robotics. A person working with automated manufacturing systems must be familiar with both fluid and electrical system basics.

An example of a hybrid fluid power system is the hoist used to lift cars in service stations, **Figure 6-26**. Both pneumatic and hydraulic systems produce the power needed to lift an automobile. Adding air under pressure to the top of a long cylinder within an oil-filled tube forces the cylinder to move upward. The tube and cylinder are normally placed in the floor so that the entire unit retracts when air pressure is removed.

Figure 6-26. A hybrid (air and oil) fluid power system drives this automobile hoist.

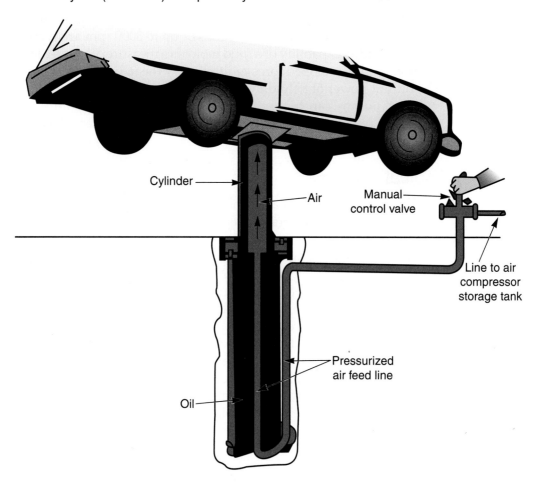

Review Questions

Write your answers on a separate sheet of paper. Do not write in this book.

1. Briefly describe the characteristics of a hydraulic system. What are some industrial applications of such systems?
2. Briefly describe the characteristics of a pneumatic system and provide some industrial applications.
3. Describe the operation of a pressure regulator.
4. Explain Pascal's law and its application to fluid power systems.
5. Define the following terms: *force, pressure, work,* and *power.*
6. Identify the function of each of the following parts of a fluid power system: motor, pump, pressure relief valve, pressure gauge, and reservoir.
7. List four types of fluid pumps. Provide a brief description of how each functions.
8. Explain the purpose of fluid conditioning in a hydraulic system. Give examples of conditioning devices used in hydraulic systems.
9. What conditioning devices are used in a pneumatic system?
10. What are some types of control devices used with fluid power systems?
11. How is direction control accomplished in a fluid power system?
12. List some types of indicators used with fluid power systems and describe the use of each.

Learning Extensions

1. Visit the National Fluid Power Association Web site at www.nfpa.com. Search the site for information about the fluid power industry and standards development. List three standards that relate to topics presented in this chapter.
2. Visit the Fluid Power Safety Institute Web site at www.fluidpowersafety.com. Review several of the safety products and write a brief summary of the use and features of one of the safety products.
3. Solve the following problem based on the variables given:
   - A cylinder with a 2 in.^{2} piston head is used to raise a 1000 lb. load. What pressure is required to accomplish this task?
   - What is the force that can be lifted using a 3 in.^{2} piston head with a pressure of 500 psi?

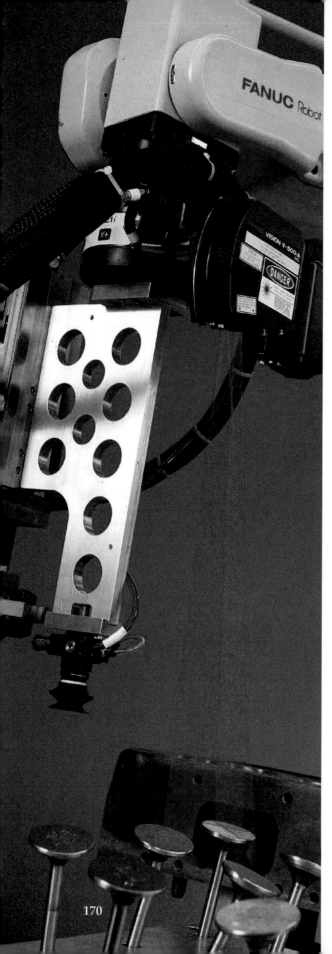

Unit III
Sensing and End-of-Arm Tooling

The control of an industrial robot often depends on the sensing system. Sensing systems use devices called transducers to convert light, heat, or mechanical energy into electrical energy. The signal output of the transducer affects the operation of the robot's end effector (end-of-arm tooling). End effectors are attached to the wrist of the manipulator and can grasp, lift, transport, maneuver, or perform operations on a workpiece.

Unit III Chapters

Chapter 7: Sensors 171
7.1 How Sensors Work 172
7.2 Types of Sensors..175
7.3 Sensor Applications...189

Chapter 8: End Effectors...................................195
8.1 End Effector Movement..................................196
8.2 Types of End Effectors....................................198
8.3 Changeable End Effectors............................204
8.4 End Effector Design..205

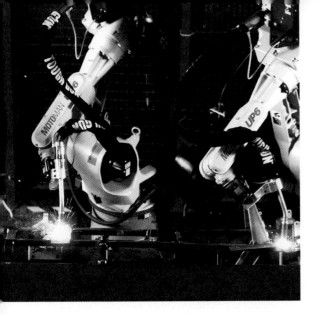

Chapter 7
Sensors

Outline

7.1 How Sensors Work
7.2 Types of Sensors
7.3 Sensor Applications

Objectives

Upon completion of this chapter, you will be able to:

- Explain the function of transducers in the operation of sensors.
- Identify the various sensors used in an automated system.
- Describe how sensors are integrated into an automated system.

Technical Terms

acoustical proximity sensor
angstrom (Å)
capacitance
capacitive transducer
computer vision sensor
eddy current proximity sensor
electromagnetic spectrum
inductance
inductive transducer
infrared sensor
laser
laser interferometric gauge
light-emitting diode (LED)
limit switch
magnetic field sensor
microswitch
nanometer (nm)
optical fibers
optical proximity sensor
opto-electronic
photoconductive device
photoemissive device
photovoltaic device
piezoelectric effect
proximity sensor
range sensor
reed switch
resistive transducer
sound sensor
speed sensing
stadimetry
strain gauge
tactile sensor
thermistor
thermocouple
thermoelectric sensor
touch-sensitive proximity sensor
transducer
triangulation
ultraviolet sensor
X-rays

Overview

The control of many industrial robots depends on various types of sensors. A sensor operates by converting light, heat, or mechanical energy into electrical energy. The resulting signal affects the operation of the machine. Sensors give robots a higher level of intelligence by improving decision-making capability. This chapter discusses some of the sensors used with robotic systems in industry.

7.1 How Sensors Work

Sensors allow robots to interact with their environment, respond to changes, and determine a course of action. Sensors help robots respond to different positions and orientation of parts, variations in the shape and dimensions of parts, and unknown obstacles in the work environment.

Transducers

Transducers convert light, heat, or mechanical energy into electrical energy. The conversion of physical quantities to electrical quantities is a basic sensing function. The signal output by the transducer is used to affect the operation of the machine. For example, a thermocouple is a transducer that converts heat energy into electrical energy. A microphone is a transducer that converts sound energy into electrical energy. Transducers may be resistive, capacitive, or inductive (**Figure 7-1**).

Resistive Transducers

Resistive transducers convert a variation in resistance into electrical variations. One type works on the potentiometer principle, **Figure 7-2**. This transducer changes resistance when the position of its movable contact is changed. Increasing the length of wire between terminals increases the resistance between those two points. Resistive transducers are often used to sense physical displacement. Displacement moves the sliding contact, which changes the resistance in the control circuit.

Capacitive Transducers

Capacitive transducers measure a change in *capacitance*, or the ratio of charge on a conductor to the potential difference between conductors. Capacitance exists when two conductive materials (plates) are separated by an insulating material. Capacitance can be increased by increasing the area of the plates or by decreasing the thickness of the insulation.

One use of capacitive transducers is sensing fluid pressure, **Figure 7-3**. In this application, the transducer is placed into the fluid line. One capacitor plate is a conductive diaphragm that senses any variation in fluid pressure. The other capacitor plate is stationary. When the fluid pressure in the line increases, the conductive diaphragm plate moves closer to the

Figure 7-1. Resistive, capacitive, and inductive sensors can be used to create electrical output that shows displacement.

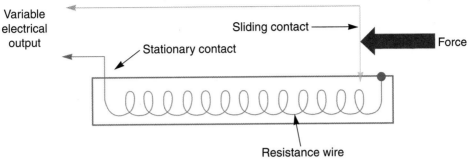

Resistive Sensor

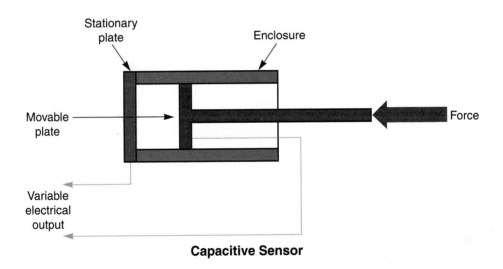

Capacitive Sensor

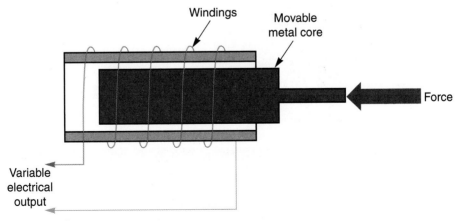

Inductive Sensor

Figure 7-2. Resistive transducers.

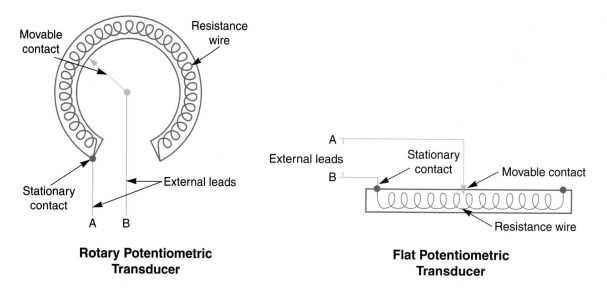

Figure 7-3. This capacitive transducer senses fluid pressure. Pressure changes cause movement of capacitor plate 1, changing the capacitance of the circuit and affecting the control signal emitted by the transducer.

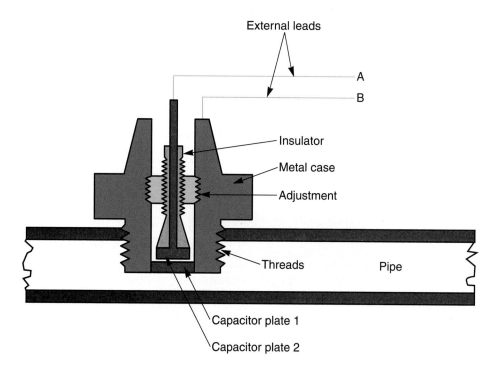

stationary plate. When the distance between capacitor plates decreases, the capacitance between the terminals increases. When fluid pressure decreases, the distance between the plates increases and capacitance decreases.

Inductive Transducers

Inductance is a property of electrical circuits caused by the magnetic field that surrounds a coil when current is flowing. In ac circuits, inductance opposes changes in current and increases as frequency increases.

An *inductive transducer* measures movement and creates signals that affect current flow, **Figure 7-4**. Inductive transducers usually have a stationary coil and a movable core. The movable core is connected to an object whose movement is to be measured. As the core changes position within the coil, the inductance of the coil varies. Current flow through the coil drops as inductance increases.

The linear variable differential transformer (LVDT) is a common type of inductive transducer, **Figure 7-5**. A movable metal core is placed in a tubular housing that has three windings. The center winding (primary) is connected to an ac source. Voltage is induced in the two outer windings by the primary winding. Initially, the voltage induced in each of the two outer windings is equal. Any movement of the core causes the voltage in one winding to increase and the voltage in the second winding to decrease. The difference between the induced voltages depends on the amount of core movement.

7.2 Types of Sensors

Many different sensing techniques and kinds of transducers are used with robotic systems. The type of sensor used depends on what is to be sensed, the accuracy required, and the environment.

Figure 7-4. A movable-core inductive transducer permits measurement of movement.

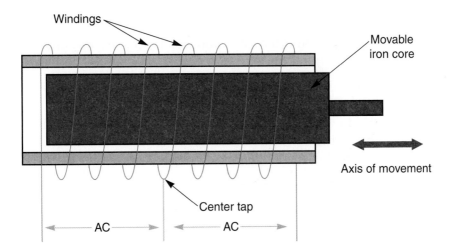

Figure 7-5. A linear variable differential transformer (LVDT) measures movement based on changes of the voltage.

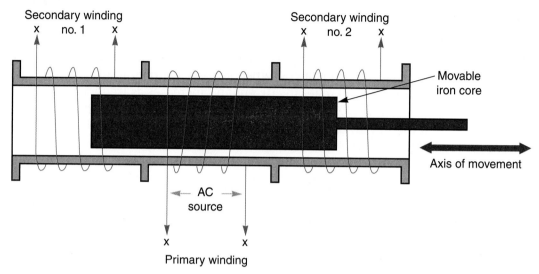

Light Sensors

Light is a visible form of radiation that occupies a narrow band of frequencies within the electromagnetic spectrum. The *electromagnetic spectrum* includes radiation frequencies for radio, television, radar, infrared radiation, visible light, ultraviolet light, X-rays, and gamma rays (**Figure 7-6**). The different types of radiation differ only in frequency or wavelength.

The human eye responds only to light in the visible frequencies, **Figure 7-7**. Each color of light has a different frequency and wavelength. Wavelengths of light are measured in *nanometers (nm)*, which is one billionth of a meter, and the *angstrom (Å)* unit, which is one-tenth of a nanometer. Visible light wavelengths range from violet (400 nm/4000 Å) to red (700 nm/7000 Å).

Light sensors respond to changes in light energy using various opto-electronic devices. *Opto-electronic* devices use a combination of optical and electronic components. These devices, along with lasers and X-ray devices, are important components in control circuits. Opto-electronic devices fall into three categories:

- *Photoemissive devices* emit electrons in the presence of light. Phototubes are a type of photoemissive device.
- *Photoconductive devices* vary in conductivity according to fluctuations in light. Their electrical resistance decreases when light is more intense and increases when light intensity decreases, **Figure 7-8**.
- *Photovoltaic devices*, or solar cells, convert light energy into electrical energy, **Figure 7-9**. When light energy falls on a photovoltaic device, it creates an electrical voltage. Although their electrical output is low, they are used with amplifying devices to drive a load.

Figure 7-6. The electromagnetic spectrum includes only a narrow band of visible light.

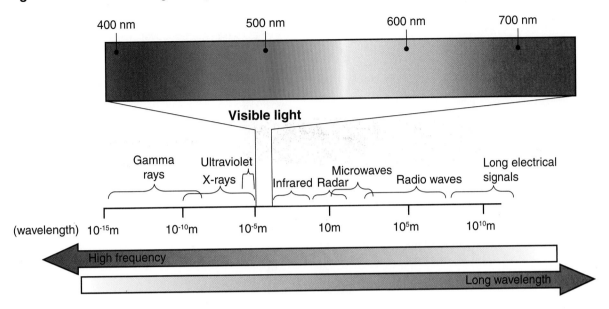

Figure 7-7. The human eye responds to visible light at wavelengths between 4000 and 7000 angstroms, with peak sensitivity in the middle wavelengths (shades of green and yellow).

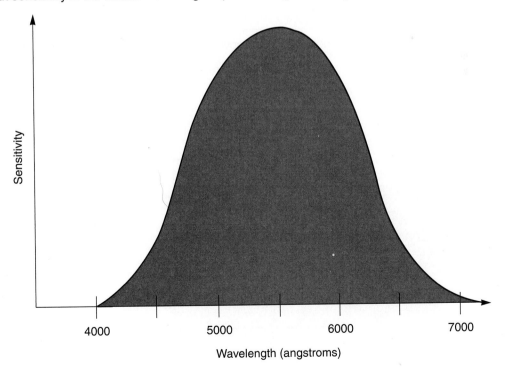

Figure 7-8. The conductivity of a cadmium sulfide cell (cutaway view) varies with the amount of light energy striking it.

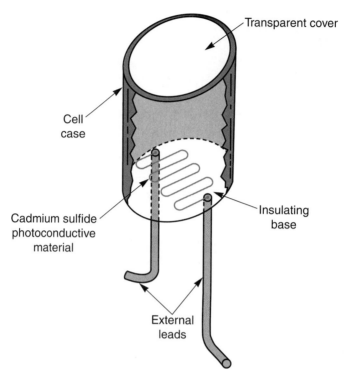

Figure 7-9. Electrons in this selenium photovoltaic cell move from layer to layer, generating electrical energy.

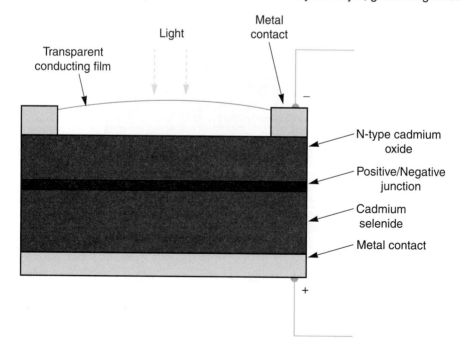

Various types of opto-electronic devices are designed to sense the position of a light beam and are used with digital control systems to produce an electrical output, **Figure 7-10**. The output is based on patterns created by the photoconductive material onto which the light beam is focused.

Infrared

Infrared sensors respond to radiation in the infrared region of the electromagnetic spectrum. All objects emit infrared thermal radiation. Infrared camera systems, **Figure 7-11**, can detect this radiation—even in darkness. Industrial uses include heat-sensitive control systems, optical pyrometers, and infrared spectroscopy for gas analysis.

Ultraviolet

Ultraviolet sensors respond to electromagnetic radiation in the ultraviolet range. Ultraviolet sensors are used for applications such as color measurement, leak detection, impurity detection, and other precision measurement applications. These sensors may also be used in machine vision applications.

Figure 7-10. This opto-electronic digital readout indicates position of a light beam.

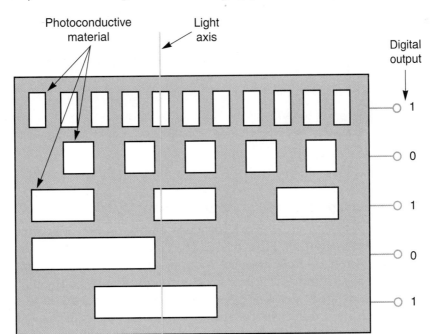

Figure 7-11. An infrared image displays the thermal radiation emitted by objects. In this image, you can see the hot liquid (red and yellow) moving through the cooler piping (blue and purple) of this processing operation. (Infrared Cameras, Inc.)

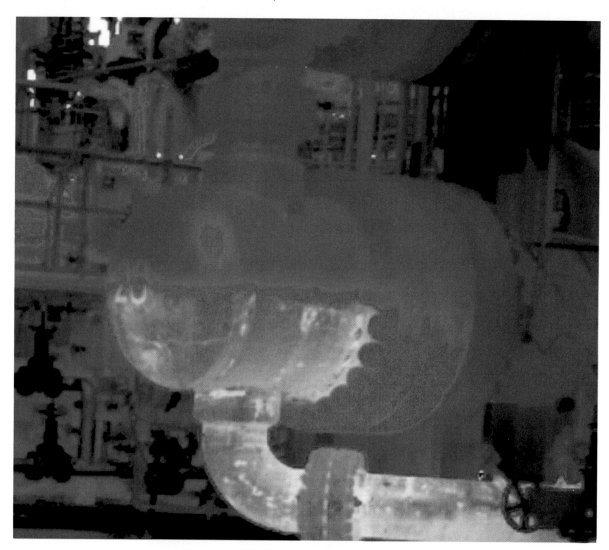

Fiber Optics

Optical fibers made of glass or plastic can transmit light from one point to another, **Figure 7-12**. The cladding (cover) material of the fibers is reflective, so light bounces from side to side within the optical fiber "light pipe" as it moves. Light travels through the fiber-optic material regardless of how the material is bent or shaped. The light can travel around corners, within a limited space, and over long distances. Fiber-optic sensors have replaced other sensor types in applications such as rotation, acceleration, magnetic field measurements, temperature, pressure, and sound.

Figure 7-12. Light can be "bent" around corners using optical fiber, sometimes called a "light pipe." A—The components of an optical fiber. B—Fiber optic strands.

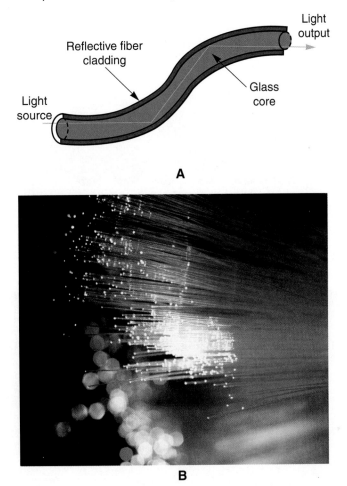

Laser

The term *laser* stands for light amplification by stimulated emission of radiation. The development of the laser had a significant and continuing impact on industrial control systems, **Figure 7-13**. A major advantage of the laser is its tightly-focused beam, which can travel long distances with relatively little spreading.

To produce a ruby laser beam, a xenon tube in the laser assembly flashes and chromium atoms in the ruby rod absorb photons of light. The chromium atoms then emit their own photons. The photons travel along the axis of the ruby rod and are reflected by the mirrors on each end, which amplifies the light. The photons "fall into step" with one another and produce only one wavelength (color) of light. The result is an intense beam of light energy.

Figure 7-13. A ruby laser produces a beam of photons that do not spread out and scatter, as ordinary light does. A—The components of a ruby laser. B—The focused beam of a ruby laser stays intact, even when reflected off of many surfaces. (© Digital Vision)

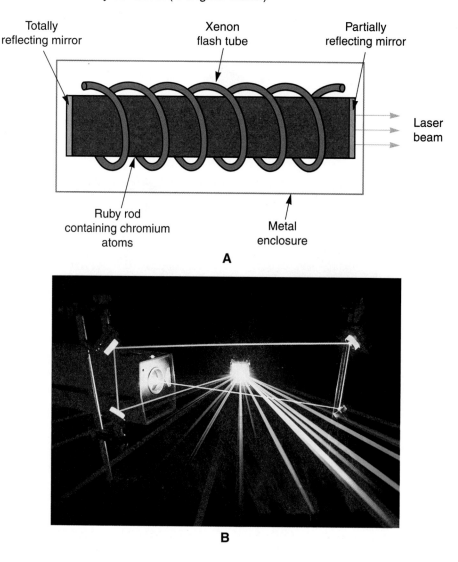

Gas Lasers. Gas lasers are often used with sensing and control systems. A common type is the helium-neon laser, **Figure 7-14**. Many other types of gas lasers are available, but all operate on basically the same principle.

A high dc potential is applied to a plasma tube using a voltage multiplier circuit and a pulse transformer. The filament within the plasma tube is heated by a 6.3v ac source. As electrons from the filament accelerate, they strike helium-neon gas atoms in the tube which causes them to ionize. The ionized gas emits light. The light reflects from a flat, fully reflective mirror at the top. The plasma tube is cut at a precise angle to control the reflection.

Figure 7-14. A laser using helium-neon gas is one of the types used in industry for sensing and control applications.

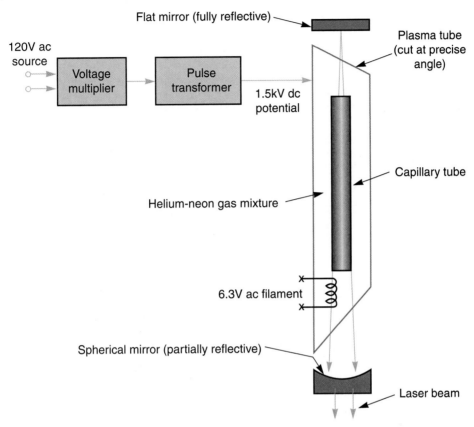

The light is reflected back and forth several times to the partially-reflective, spherical mirror at the bottom and is concentrated into a laser beam that is emitted through the spherical mirror.

Semiconductor Lasers. It is also possible to generate laser beams using semiconductors. Semiconductor injection lasers are very efficient and extremely tiny in size compared to other lasers, **Figure 7-15**. These lasers have a resonant cavity similar to other lasers, except the cavity is formed on the chip of semiconductor material—typically gallium arsenide chips. The end faces of a gallium arsenide chip are parallel and flat. Since gallium arsenide is a reflective material, mirrors are not needed for the laser. This is a distinct advantage in terms of complexity and cost.

As current flows through the chip, light is emitted from the gallium arsenide. Atoms collide in the area where positive and negative semiconductor materials meet, releasing additional photons. Due to the reflective properties of the gallium arsenide, a wave of photons is developed between its flat surfaces. The back-and-forth movement of this wave creates the resonant action required to produce a laser beam.

Figure 7-15. This semiconductor injection laser requires no mirrors. It is tiny and very efficient.

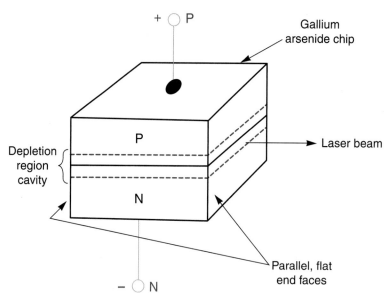

X-rays

X-rays are invisible rays within the electromagnetic spectrum between ultraviolet and gamma rays. The short wavelength of X-rays allow them to pass through the human body—a property that makes them valuable in medical treatment and analysis.

It is possible to use a vacuum tube to produce rays similar to those emitted by radium, **Figure 7-16**. The cathode in an X-ray tube is heated using a filament voltage. The anode is constructed of a heavy metal. A high positive potential is applied to accelerate the electrons emitted from the cathode at a very rapid rate. The electrons strike the anode with such velocity that X-rays are created. If the potential is increased, the frequency of the X-rays increases and the wavelength decreases.

X-ray tubes can operate with applied potential in excess of one million volts dc. The resulting X-rays are similar to the high-frequency gamma rays emitted by radium. In industry, X-rays are used to control processes that involve metals. The short wavelength of X-rays allows them to pass through metals and reveal structural characteristics.

Vision Sensors

Vision sensors may be used to recognize objects or to measure the characteristics of objects. Camera-equipped computer systems are able to identify a specific part using a video camera and can distinguish that particular part from any other. An object may be identified by its shape, outline, or area, regardless of its orientation to the camera. See **Figure 7-17**.

Figure 7-16. The components of an X-ray tube.

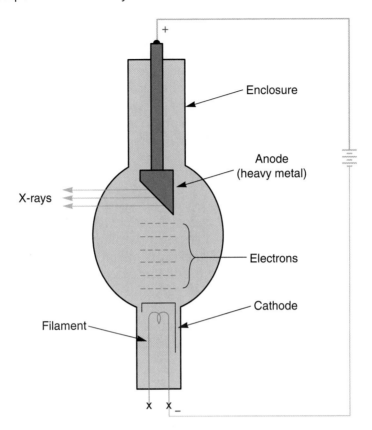

Figure 7-17. A laser is used to check the quality of this weld. (Schlumberger)

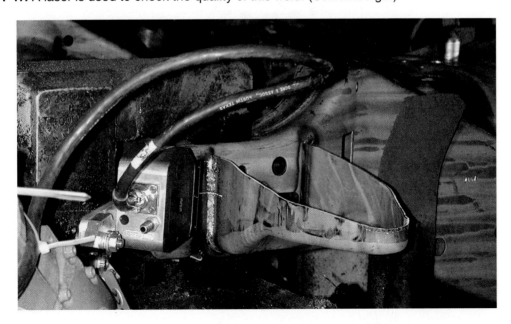

Computer vision sensors detect spatial relationships and provide depth information using stadimetry and triangulation. *Stadimetry* determines the distance to an object based on the apparent size of the camera image. *Triangulation* involves measuring angles and the base line of a triangle to determine the position of an object.

Position detection can be accomplished by placing a camera on the robot's end effector. The feedback is used to guide the end effector to a specific location. This process is referred to as *visual servoing*, and is used to move material from one spot to another. Servo movement may be used with either stationary or moving objects.

Sound Sensors

Sound sensors rely on the piezoelectric effect to convert sound to electrical energy. The *piezoelectric effect* occurs when certain crystals are subjected to mechanical stress and an electrical potential is developed in them. Rochelle salt and quartz are examples of these crystals.

A common application of the piezoelectric effect is the operation of crystal microphones. When sound waves (vibrations in air) strike a piezoelectric crystal, **Figure 7-18**, a voltage develops across the crystal. The voltage is amplified by the control system. In this way, mechanical energy (vibration) is converted to electrical energy.

Figure 7-18. One common application of the piezoelectric effect is crystal microphones.

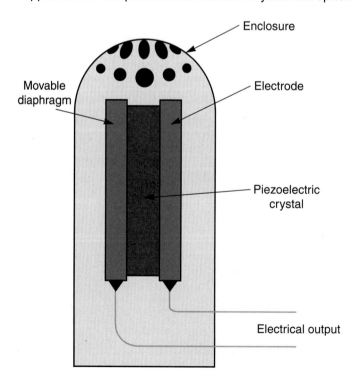

Temperature Sensors

Thermoelectric sensors produce a change in electrical output due to a fluctuation in temperature. They are used to control processes in which temperature must be held at a given level, or must not exceed a specific point.

One device that is commonly used for heat sensing is the *thermistor*. The resistance of this temperature-sensitive resistor decreases as temperature increases (and vice-versa). This phenomenon is referred to as a "negative temperature coefficient of resistance." Various metal-oxide semiconductor materials are used to construct thermistors, **Figure 7-19**. Thermistors are manufactured in a wide range of resistance characteristics and temperature coefficients.

Figure 7-19. In a thermistor, a decrease in temperature causes increased resistance. These are some of the common forms of the device.

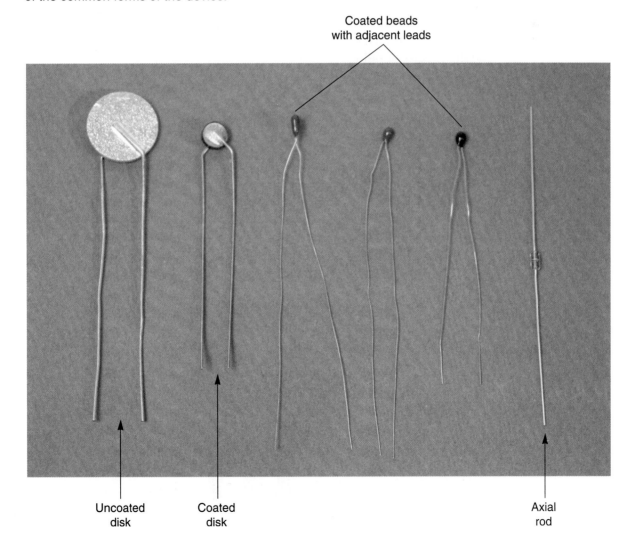

Thermocouples are devices that convert heat energy into electrical energy. A thermocouple, **Figure 7-20**, consists of two dissimilar metal strips fused together at one end. The metals are usually a combination of iron-constantan, copper-constantan, and platinum-rhodium. Different metal combinations respond to different ranges of temperature. When the fused end is heated, a voltage develops at the ends that are not connected. This voltage exists because of the differing coefficients of expansion in the two metals. The voltage produced by a thermocouple is usually in the millivolt range.

Magnetic Field Sensors

Magnetic field sensors identify a change in an existing magnetic field without making physical contact with objects in the environment. This sensor detects changes in the magnetic field and sends a signal indicating the disturbance. Magnetic field sensors can be used in positioning operations, control tasks, range detection, and many other applications. Magnetic field sensors used in industrial applications typically use a permanent magnet as the constant magnetic field. Permanent magnets are not as susceptible to interference from other devices in the work area or from the Earth's magnetic field.

Reed Switches

Reed switches are simple magnetic field sensors that either make or break contact in response to changes in a magnetic field, **Figure 7-21**. Two flat metal strips, or "reeds," are housed in a hollow glass tube filled with an inert gas. When the reeds are exposed to a magnetic field, they are forced together, completing an electrical circuit. When the magnetic field is turned off, the reeds spring open and the circuit is broken. The reeds are housed inside a hermetically sealed glass tube, so contact sparks are isolated from the outside. The reeds are also isolated from outside dust and corrosion, which improves the life expectancy of the switch.

Figure 7-20. A thermocouple converts heat energy into electrical energy.

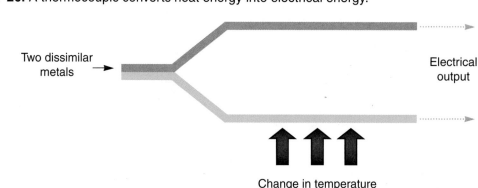

Figure 7-21. A reed switch responds to changes in a magnetic field.

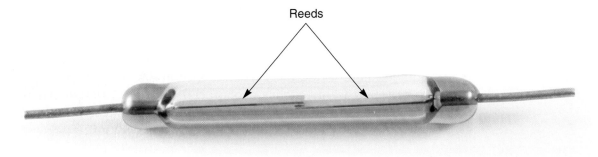

7.3 Sensor Applications

There is a wide variety of sensor types available and an equally wide variety of industrial and robotic applications for sensors. Whether triggered by light, magnetic disturbance, or sound, sensors are an integral component of robotic operations and control.

Speed Sensors

Speed sensing involves measuring the rotary motion of shafts, gears, pulleys, and other rotating components of industrial equipment. A dc tachometer system is one method used for speed sensing, **Figure 7-22**. The tachometer is connected directly to a rotating piece of equipment. The rotation turns the shaft of the small dc generator, creating voltage. The

Figure 7-22. A dc tachometer senses speed changes and displays them as meter fluctuations.

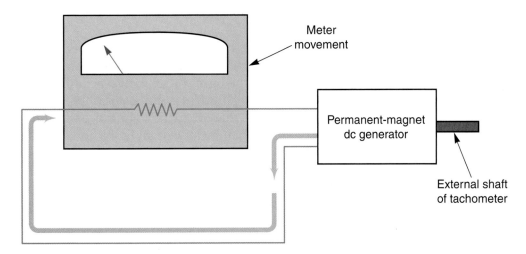

generator's voltage output is directly related to the rotation rate of the shaft. Voltage output is translated into speed readings, which are used to control equipment operation.

Another method of speed sensing is an electronic tachometer, which offers precision and ease of use. A reflective material is placed on the surface of the rotating portion of the equipment. Light emitted from the tachometer is reflected back when it encounters the reflective material. The reflected light reaches a photocell on the tachometer, which converts the pulses of reflected light energy into electrical signals. The electrical signals are used to measure the speed of rotation.

Mechanical Movement Sensors

Mechanical movement is a change in dimension resulting from an applied force. This change can be sensed using a *strain gauge*, **Figure 7-23**. A strain gauge is made of fine-gage resistance wire, about 0.001″ (0.025 mm) in diameter, and is mounted on a strip of insulation. The wire is flexible, so it stretches when subjected to stress. As the wire stretches, its cross-sectional area is reduced and its resistance changes. When used in a robotic system, strain gauges emit electrical signals (feedback) based on the amount of pressure the "fingers" of a robot exert to lift an object.

Semiconductors are commonly used in place of traditional metal strain gauges. The rate of change in resistance is approximately 50 times higher than metal strain gauges, and semiconductors are more sensitive to small changes. A semiconductor strain gauge is as stable as the metal type, in terms of consistency of signal changes caused by resistance variation. However, semiconductor strain gauges typically produce a higher electrical signal (feedback voltage) output.

Proximity Sensors

Most *proximity sensors* detect either the absence of an object or the presence of an object within a certain distance. Some proximity sensors provide feedback about the distance between the sensor and an object, such as an end effector.

Optical proximity sensors measure the amount of light reflected from an object and can respond to either visible or infrared light. Incandescent lights can be used as the light source, however LEDs are generally preferred. *LEDs (light-emitting diodes)* (solid-state lamps) are small, lightweight opto-electronic devices that are made to produce different colors of light. LEDs are more reliable than incandescent lights and are not sensitive to shock and vibration. The semiconductors used in LEDs produce light when an electric current is applied. The response varies according to the type of semiconductor used. The wavelength of radiated energy from an infrared LED is beyond the visible range and, therefore, cannot be seen by the human eye.

Figure 7-23. A strain gauge responds to mechanical energy by changing its resistance. A—The components of a strain gauge. B—Resistance increases when the strain gauge is under tension. C—Resistance decreases when the strain gauge is under compression.

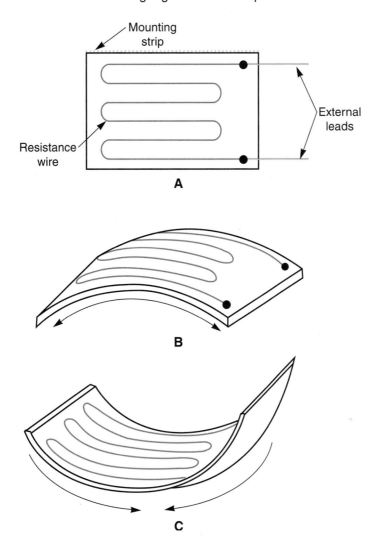

Eddy current proximity sensors produce a magnetic field in the small space of a detector unit, which can be mounted in a probe. The magnetic field induces eddy currents into any conductive material that is near the probe. A pick-up coil senses a change in magnetic field intensity when an object enters the field.

Acoustical proximity sensors react to sound. Standing sound waves are generated within a cylindrical, open-ended cavity inside the sensor. The presence of a nearby object interferes with these waves, which alters the sensor's output. A microphone may be used to detect a change in sound pressure and measure the distance of the object from the sensor.

Touch-sensitive proximity sensors operate on capacitance developed by a large conductive object (such as the human body). This capacitance changes the frequency of an electronic circuit. A conductive plate or rod may be used to sense contact.

Tactile sensors indicate the presence of an object by touch, while stress sensors produce a signal that indicates the magnitude of the contact made. A simple type of touch sensor is a microswitch. *Microswitches* are electrical switches that are turned on or off with a very small amount of force. They are small, durable, and easy to activate. Common applications for microswitches include computer mouse buttons and joystick controls. Limit switches also respond to contact with an object. *Limit switches* have an actuator that is mechanically linked to a set of contacts, which are either normally open or normally closed (**Figure 7-24**). These switches may be used with robotic systems to sense the presence (or absence) of an object. When activated, contacts that are normally open are closed and contacts that are normally closed are opened. This causes a change of state to an electrical circuit. Limit switches are rugged, simple, easy to install, and reliable. They are used for more heavy-duty applications than microswitches, including conveyor systems, milling, and drilling machinery operations.

Range Sensors

Range sensors determine the precise distance from the sensor to an object. Such devices are useful for locating objects near a work station or for controlling a manipulator. A *laser interferometric gauge* is a range sensing system that is sensitive to humidity, temperature, and vibration. Some range sensing systems use a television camera that operates with sonar (sound).

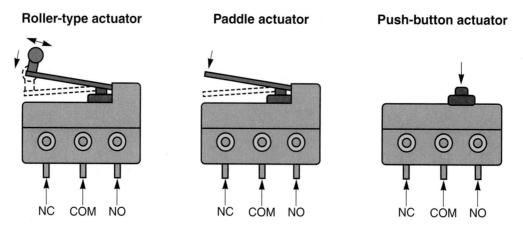

Figure 7-24. Types of limit switches. NC = Normally closed contact, NO = Normally open contact, COM = Common.

Inventor Spotlight: Nick Holonyak, Jr.

Nick Holonyak, Jr.

Light-emitting diodes (LEDs) can be found in such everyday objects as alarm clocks, flashlights, and children's toys; but are also used in complex systems like infrared sensors and fiber-optic communication technology. These tiny devices were invented by Nick Holonyak, Jr. in 1962 while working in the Advanced Semiconductor Laboratory at the General Electric Company. Though the first visible LED emitted red light, they are now available in an array of colors.

Nick Holonyak has been involved in many other inventions and developments that have made an immediate impact on technology and modern society.

- 1958: Helped develop the first light dimmer switch while working at General Electric.

- 1960: Created the red-light semiconductor laser (laser diode), which is used in CD players, DVD players, and CD-ROM drives.

- 1990: Developed the technology that spawned the creation of the lasers used in copy machines, laser printers, computer mice, and communication and data transmission.

Nick Holonyak is also the co-inventor of the infrared light-emitting transistor (LET). This recent development has the potential to revolutionize telecommunication systems and computer systems, in terms of both size and capability.

Review Questions

Write your answers on a separate sheet of paper. Do not write in this book.

1. What is the purpose of equipping a robot with a sensing system?
2. What is a transducer? Identify three types.
3. What are three categories of opto-electronic devices? Define each category and give examples of devices in each.
4. What is the difference between an infrared sensor and an ultraviolet sensor?
5. Describe two types of laser sensors.
6. How are X-rays used in industrial control?
7. How can you use a sensor to measure depth?
8. What is visual servoing?
9. Explain the piezoelectric effect.
10. Discuss two types of temperature sensors.
11. Describe how reed switches operate.
12. How is a strain gauge used to sense mechanical movement?
13. What are tactile sensors? Give an example of how tactile sensors are used.

Learning Extensions

1. Make a list of five different types of sensors. Search magazines or the Internet for examples of how the sensors are used in the robotics industry. Identify three different applications for each of the sensor types you listed.
2. For each of the sensors identified in the previous activity, note the type of transducer used by the sensor (resistive, capacitive, or inductive).

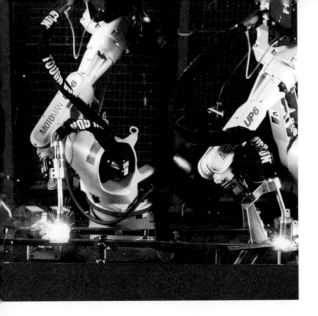

Chapter 8
End Effectors

Outline

8.1 End Effector Movement
8.2 Types of End Effectors
8.3 Changeable End Effectors
8.4 End Effector Design

Objectives

Upon completion of this chapter, you will be able to:

- Discuss the similarities and differences between the movement of an end effector and the human hand.
- Explain the operation of various types of grippers used in robotic applications.
- Describe the difference between end effector grippers and end effector tools.
- Identify the benefits of changeable end effectors.
- List important factors and desirable characteristics to be considered in the design of end effectors.

Technical Terms

automatic tool changer
collet gripper
compliance
cylindrical grip
electromechanical gripper
expandable gripper
gripper
hook movement
lateral grip
magnetic gripper
mechanical finger gripper
nonprehensile movement
oppositional grip
overload sensor
palmar grip
prehensile movement
remote-center compliance (RCC) device
spherical grip
spread movement
tool
vacuum gripper

Overview

A key component in robot design is the end effector, or end-of-arm tooling. End effectors are devices attached to the wrist of a manipulator. They can grasp, lift, transport, maneuver, or perform operations on a workpiece. This chapter discusses the two types of end effectors—grippers and tools—and factors that influence end effector design.

8.1 End Effector Movement

The robot's end effector is a clumsy imitation of the human hand. The human hand has the ability to adjust, grasp, pick up, and rotate different objects. It also has intuitive sensing capabilities, which allow the hand to adjust to imperfections in objects. The human hand can perform both prehensile and nonprehensile movements.

Prehensile Movements

Prehensile movements require the use of the thumb to grasp objects. Curled fingers and the opposing thumb provide the hand with the dexterity needed for these movements. The opposed thumb is very important because it enables humans to pick up and manipulate very small objects.

As a child grows and coordination develops, so does the ability to use various grips, **Figure 8-1**. The hand makes the following five basic prehensile (gripping) movements:

- *Palmar grip*. Wrapping the fingers and thumb around an object, such as a baby's rattle or a human finger, to grasp.
- *Cylindrical grip*. Forming a "C" shape with the fingers and thumb to grasp a cylindrical object, such as a drinking glass or water bottle.
- *Spherical grip*. Using the fingers to hold round objects, such as holding and throwing a baseball.
- *Lateral grip*. Grasping flat objects from the sides with the fingers and thumb, rather than around.
- *Oppositional grip*. Using the tip of the index finger and thumb to hold an object.

Nonprehensile Movements

Nonprehensile movements do not require particular finger dexterity or use of the opposed thumb. These movements include pushing, poking, punching, and hooking. Two specific types of nonprehensile movements are hook and spread, **Figure 8-2**. The *hook movement* involves curling the tips of the fingers to pull or lift objects. To perform the *spread movement*, the fingers and thumb are extended outward until they make contact with the interior walls of a hollow object. The force of the fingers against the walls of the object allows it to be picked up and carried.

Figure 8-1. The human hand is capable of five basic prehensile grips—palmar, cylindrical, spherical, lateral, and oppositional.

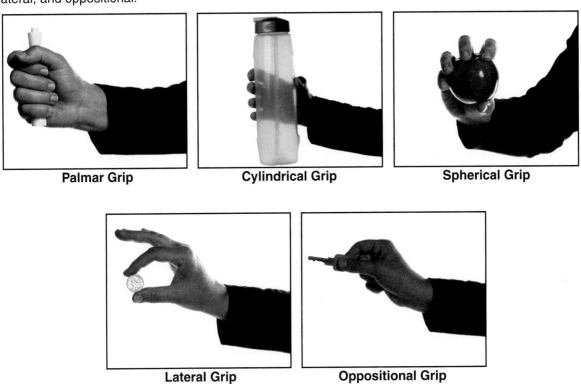

Figure 8-2. Nonprehensile movements, such as hook and spread, do not depend on movement of the thumb.

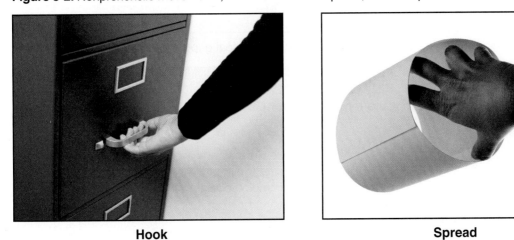

8.2 Types of End Effectors

End effectors can be classified as grippers or tools. *Grippers* are end effectors that perform prehensile movements by grasping objects and moving them. *Tools* are end effectors that execute nonprehensile movements to perform specific tasks, such as welding or painting.

Grippers

Robots use a variety of grippers to grasp, handle, and transport parts. Some common types of grippers are presented in the sections that follow, **Figure 8-3**.

Mechanical Finger Grippers

Mechanical finger grippers are the type most commonly used for grasping objects. Mechanical finger grippers are generally used for grasping parts within a confined space, reaching into channels, or picking and placing any object that has a simple shape. The fingers typically move in a parallel or angular motion to grasp an object, **Figure 8-4**. The gripper's fingers are opened and closed using mechanical linkages, gears, cables, chains, or pneumatic actuators.

Two-finger Grippers. These grippers have two stiff fingers that simulate the motions of the human thumb and index finger (**Figure 8-5**); one or both of the fingers may move. For an additional degree of freedom and more flexibility during operations, a two-finger gripper may be equipped with a rotating joint (**Figure 8-6**).

To handle objects with different shapes, fingers can be interchanged. For grasping cylindrical objects, V-shaped fingers are recommended (**Figure 8-7**).

Figure 8-3. This classification scheme can be used for mechanical grippers.

Gripper End Effectors			
Gripper Type	**Gripper Configuration**	**Gripper Movement**	**Internal/External Gripping**
Mechanical finger	Two-finger Three-finger Four-finger	Parallel or angular	Internal and external
Collet	Round Square Hexagonal	360° clamping contact	Internal and external
Vacuum	One or more suction cups	Vacuum/suction	External
Electromechanical	Permanent magnet Electromagnet	Magnetic attraction	External

Figure 8-4. Two types of motion are made by mechanical grippers—parallel and angular.

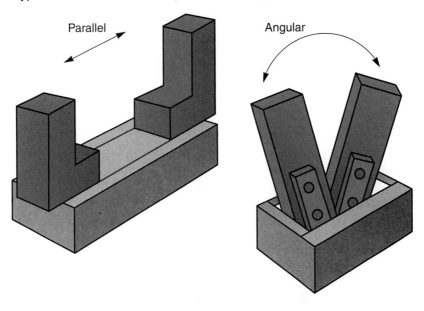

Figure 8-5. This two-finger internal gripper is hydraulically operated with parallel motion and is designed to grasp heavier loads. (SCHUNK Intec Inc.)

Figure 8-6. A rotary joint gives this gripper one additional degree of freedom. (PHD, Inc.)

Figure 8-7. This two-finger gripper has two points of contact per finger, which allows it to securely grasp large cylindrical objects. (Schunk-USA)

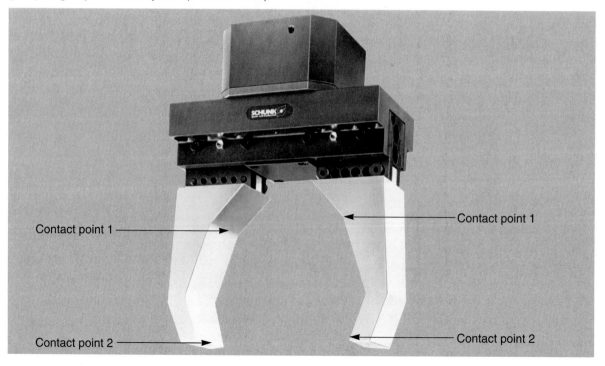

The "V" shape has two points of contact on each finger, ensuring the object is centered. Self-aligning, padded fingers are used to grip flat objects. Fingers may also have cavities of more than one size or shape. Multi-cavity fingers may be necessary if an object changes shape or size during processing.

Three-finger Grippers. Three-finger grippers simulate the action of the human thumb, index finger, and third finger, **Figure 8-8**. This configuration is better than two-finger grippers for grasping curved, spherical, or cylindrical workpieces.

Four-finger Grippers. Four-finger grippers can grasp square and rectangular parts. The opposing fingers close simultaneously to permit easier part orientation. Four-finger grippers are often constructed from a pair of two-finger grippers.

Collet Grippers

Collet grippers are used to pick and place cylindrical parts that are uniform in size. Unlike finger grippers, collet grippers deliver 360 degrees of clamping contact. They have a strong clamping force for rapid part transfer and are also used for grinding and deburring operations. Collet grippers are typically controlled by a solenoid and have excellent repeatability. They are available in round, square, or hexagonal shapes.

Figure 8-8. Three-finger grippers can grasp and handle objects of various shapes. (SCHUNK GmbH & Co. KG)

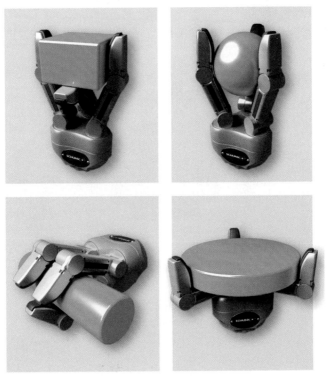

Vacuum Grippers

Vacuum grippers consist of one or more suction cups made of natural or synthetic rubber and are extremely lightweight and simple in construction. The number, size, and type of suction cups used depends on the weight, size, shape, and type of material being handled. Multi-cup vacuum grippers increase the contact surface area, which allows workpieces of increased size and weight to be handled (**Figure 8-9**). Vacuum grippers can be used on curved and contoured surfaces, as well as flat surfaces.

The flexibility of the suction cups provides the robot with a certain amount of adaptability. Therefore, exact positioning is not as critical as with some other types of grippers. To allow for unevenness in a part's surface, some vacuum cups are spring-loaded or mounted on a ball joint.

Electromechanical Grippers

Electromechanical grippers, also called *magnetic grippers*, use a magnetic field created by a permanent magnet or an electromagnet to pick up an object. Objects that have flat, smooth, clean surfaces are the easiest to handle.

Figure 8-9. When an object with large surface area must be grasped, a multi-cup vacuum gripper is often used. The gripper pictured also has curved fingers to provide additional support when moving large objects. (Pacific Robotics, Inc.)

Grippers that operate using a permanent magnet are well-suited for explosive environments, because they do not require an electric power source that could spark. Once a part is moved, it is released from the gripper by exerting force to pull it away from the magnetic field.

An electromagnetic gripper is energized by a dc power source. An object is released from the gripper when the power source is interrupted. To speed up release time, the current is not cut off. Instead, the direction of current flow is momentarily reversed.

There are certain disadvantages associated with using electromechanical grippers. These grippers can only be used to handle materials containing iron, which limits possible applications and locations. Metal shavings and other small metal particles are attracted to the magnet when parts are machined. If these metal particles accumulate on the magnet, they can scratch the surface of a part or cause misalignment of parts. The temperature of a workpiece must also be considered. The effectiveness of magnetic force declines when workpieces are heated to several hundred degrees, as may be the case in some processing operations.

Electromechanical grippers do, however, have certain advantages over vacuum grippers. An electromechanical gripper has a longer life and can handle hotter and heavier objects. Also, electromechanical grippers immediately grip a part, while vacuum grippers require time to build up the necessary pressure. Electromechanical grippers are custom designed for specific applications—few are available as off-the-shelf items.

Tools

A robot's arm can be equipped with various types of tools to perform specific tasks. Some of the common tools used on robots today are spot welding guns, inert gas arc welders, stud welders, gluing guns, spray guns, drills, milling heads, deburrers, polishers, pneumatic screwdrivers, and nut-drivers.

The tools used for robotic applications may be categorized as:

- welding tools
- material application tools
- machining and assembly tools

Welding Tools

Welding tools vary according to the type of welding operation performed. Welding robots have tool changers that provide flexibility, reduce equipment downtime, and improve productivity. One representative type of welding tool is an arc welding torch attached to a robot arm. For this application, a collision sensor could be used to attach the welding torch to the robot arm.

Spot welding operations can also be performed with robotic systems. Tool changers allow a spot weld device to be attached to a robot arm so multiple functions can be performed within a work cell. In addition, laser and gun welding systems may be utilized with the proper tool changers.

Material Application Tools

Material application tools are also used in palletizing, packaging, and handling operations. Industries are very sensitive to productivity, so tool changers must be efficient to avoid equipment downtime. Rapid tool changing during material handling automates and improves the process.

Machining and Assembly Tools

Tooling operations performed for machining and assembly use robots to perform multiple tasks that require several different types of tools. Rapid tool changing is critical to productivity for these operations, as is the repeatability of every production cycle. Robots can perform tooling quickly with tool changers that connect the robot arm to the necessary tool in only a few seconds. Collision sensors are attached to the robot arm to protect equipment from being damaged during operation.

8.3 Changeable End Effectors

Robots used in manufacturing require either multipurpose or changeable tooling. This type of tooling increases the effectiveness and flexibility of a robotic system. Equipping the manipulator with a multifunction end effector allows a single end effector to perform multiple tasks in a single process or on a single workpiece. Another option is to design an easily changeable end effector, or quick-change tooling.

Quick-change tooling offers several end effector options that can be readily attached and removed from the manipulator, as the process requires. With quick-change tooling capabilities, a robot can handle parts of various shapes and perform a wider range of assembly tasks and machining operations. Adding this flexibility in the robotic system increases overall productivity.

When designing a robotic system with changeable end effectors, two considerations must be addressed—interface adapters must be standardized and operators must be able to change end effectors with minimum downtime. To make interface adapters standard for use with many different end effectors, they must have compatible mounting and securing systems. All the connectors for electric, hydraulic, and pneumatic components should be provided with the adapters. Since robots come in different sizes, different sized standard adapters may need to be provided, as well. Automatic tool changers allow end effectors to be changed with increased efficiency. *Automatic tool changers* are equipped with several different end effectors and can automatically change an end effector when necessary, **Figure 8-10**. This automated option reduces the downtime required to change the end effector and decreases the risk of operator injury when a manual change is performed.

Figure 8-10. Typical automatic tool changer applications. A—An automatic tool changing system permits this robot to execute three operations automatically: cleaning, primer coating, and gluing. (Schunk-USA) B—A large gantry robot is equipped with an automatic tool changing system. Note the extra-long fingers on the gripper allowing the clamping of large rings. (Schunk-USA)

A   B

8.4 End Effector Design

To determine the type of end effector needed to do a job, a study must be performed to evaluate the operation, the workpiece(s), and the environment. The end effector may be subjected to extreme temperatures or make contact with abrasive or corrosive materials. Special protective materials and shielding devices may be necessary to protect the manipulator.

Objects to be moved may vary in shape, size, and weight. The workpieces may also change in shape, size, or weight *during* a given process. The end effector must be able to adjust to such changes. Other conditions to be considered are the fragility of the workpiece, the surface finish, and the type of material used to construct the workpiece. For example, if an object is made of ferrous material (one that contains iron), a magnetic gripper may be used.

Additionally, problems involving inertia, center of mass, gripping force, or friction between the part and the gripper may need to be addressed. Other concerns might involve part orientation, gripper sensing capabilities, or interaction with other equipment.

Desirable Characteristics

The end effector should have the strength necessary to carry out its tasks and withstand rigorous use. It should also be equipped with measures to guard against damage from strain. A breakaway device should be installed to prevent damage to the robot's arm or wrist if the end effector becomes stuck. End effectors that use friction to hold objects are not typically prone to damage caused by excessive strain because the grasped object will slip out of the gripper when an opposing force is applied. However, an opposing force of this type could cause joint slippage and alter the accuracy of the robot's positioning. *Overload sensors* detect obstructions or overload conditions within fractions of a second. The controller shuts down the robot before damage occurs. Unlike breakaway joints, overload sensors do not have parts to be replaced after it operates and do not require reprogramming.

Another desirable characteristic is compliance. *Compliance* is the ability to tolerate misalignment of mating parts. For the assembly of close-fitting parts, this characteristic is essential. Compliance prevents the part from jamming, wedging, and wearing. Some end effectors have a certain amount of compliance built into their design. End effectors that lack this capability may use a *remote center compliance (RCC) device*. This device is installed in the wrist of the robot and helps to compensate for workpiece misalignment or irregularities. Robots equipped with an RCC device can perform precise tasks such as inserting bearings into a housing with clearance of only 0.0005" (0.013 mm).

Custom-Designed End Effectors

In addition to standard grippers and tools, custom end effectors can be designed for a particular application, **Figure 8-11**. Custom-designed end effectors broaden the range of tasks that robotic equipment is capable of performing.

Grippers that are designed to handle fragile objects are an example of custom-designed end effectors. Special soft, flexible cups have been developed for delicate applications, such as handling light bulbs and eggs. Custom-designed end effectors may combine gripper types or be uniquely shaped to meet the needs of a particular operation, **Figure 8-12**.

Expandable Grippers. *Expandable grippers* clamp irregularly shaped workpieces using mechanical fingers equipped with hollow rubber envelopes that expand when pressurized. The envelopes ensure even distribution of surface pressure. One type of expandable gripper surrounds an object, gripping it from the outside. The other type grips hollow objects from the inside. Expandable grippers are ideal for handling fragile parts or parts that vary a great deal in size.

Figure 8-11. Custom-designed end effectors. A—The gripper on this clean-room robot transfers large silicon wafers from one container to another. (PRI-Precision Robots, Inc.) B—This three-fingered, internal gripping end effector is used in a production line for machining precision gears. (Schunk-USA) C—This dual end effector has internal and external gripping capabilities. (Schunk-USA) D—A parallel-motion gripper is used to insert components into a printed circuit board prior to soldering. (Schunk-USA)

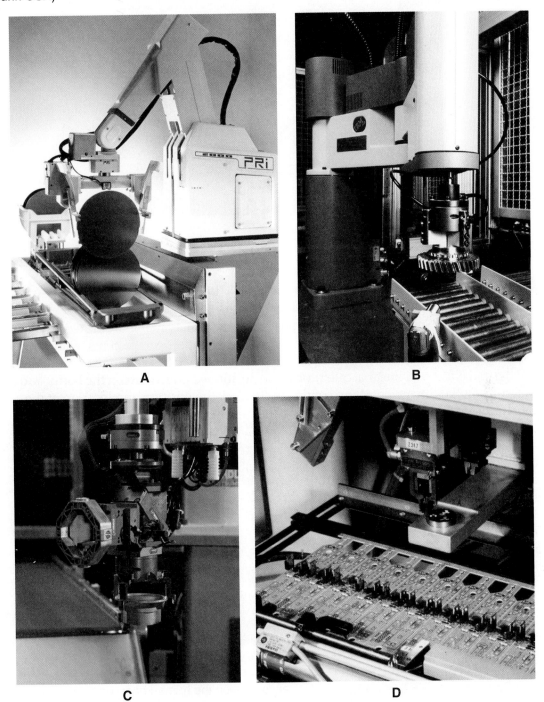

Figure 8-12. This gripper is equipped with a vision guidance system that allows the robot to find objects that are not placed symmetrically. This type of device can also be used in environments with changing conditions, such as lighting levels, and for locating and gripping objects that are placed randomly. (De-STA-Co)

Robotics in Society: CyberKnife® System

The CyberKnife System is a robotic radiosurgery system available from Accuray™ Incorporated. This system offers a non-invasive treatment option for patients with cancerous tumors. The manipulator of the CyberKnife System has an extensive range of motion, which allows the end effector (a linear accelerator) to be positioned anywhere on the body. Additionally, high-resolution image detectors continuously track the position and location of the target tissue during the treatment session. This information is communicated to the manipulator, which automatically adjusts the end effector to compensate for movement of the patient or tissue. The continuous tracking capabilities of the CyberKnife System provide more accurate treatment and reduce damage from the radiation beams to surrounding tissue.

Traditional radiosurgery systems are confined to treating tumors in the head and neck. Accuray made use of new technologies to develop a flexible and precise robotic radiosurgery system that offers a much wider range of patients an effective option in their fight against cancer.

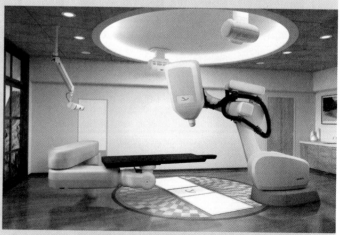

The end effector of the CyberKnife System is a linear accelerator that is used to treat tumors anywhere on the human body. (Image used with permission from Accuray Incorporated.)

Review Questions

Write your answers on a separate sheet of paper. Do not write in this book.

1. Distinguish between prehensile and nonprehensile movements.
2. In what important ways do end effectors differ from the human hand?
3. What are two major classifications for end effectors?
4. What kinds of tasks can be accomplished using a three-finger gripper?
5. Explain how a vacuum gripper operates.
6. What is a major disadvantage of an electromechanical gripper?
7. How does changeable tooling impact the operation of a robotic system?
8. What are the advantages of using automatic tool changers?
9. Identify several design considerations that apply to end effectors.
10. Explain the difference between breakaway devices and overload sensors.
11. What is compliance? How can the compliance of an end effector be increased?

Learning Extensions

1. Pay particular attention to the movements of your hands in the course of a day. Make note of common movements by identifying them as either prehensile or nonprehensile, naming the grip or movement, and listing the task you performed with the hand movement.
2. Conduct a Web search to identify end effector manufacturers. Develop a list of the types of end effectors you find and categorize them into common types.

Unit IV
Control Systems and Maintenance

The heart of a robotic control system is a microprocessor linked to input/output and monitoring devices. The control system has a series of instructions (a program) stored in its memory. The program supplies the commands that control motors and hydraulic or pneumatic systems to activate the robot's motion control mechanism. The motion control mechanism is typically an actuator, which is a device that converts power into robot movement.

Unit IV Chapters

Chapter 9: Computer Systems and Digital Electronics 211
9.1 Computer Systems 212
9.2 Digital Number Systems 219
9.3 Binary Logic Circuits 226
9.4 Computer Programming 236

Chapter 10: Interfacing and Vision Systems ... 241
10.1 Interfacing 242
10.2 Machine Vision 247

Chapter 11: Maintaining Robotic Systems 253
11.1 Troubleshooting 254
11.2 General Servicing Techniques 256
11.3 Preventative Maintenance 261

Chapter 12: Robots in Modern Manufacturing 265
12.1 Using Robots in Manufacturing 266
12.2 Evaluating Potential Uses for Robots 269
12.3 Preparing an Implementation Plan 275

Chapter 13: The Future of Robotics 281
13.1 Fully-Automated Factories 282
13.2 Robots Outside the Factory 282
13.3 Artificial Intelligence (AI) and Expert
      Systems 285
13.4 Impacts on Society 287
13.5 Your Future in Robotics 289

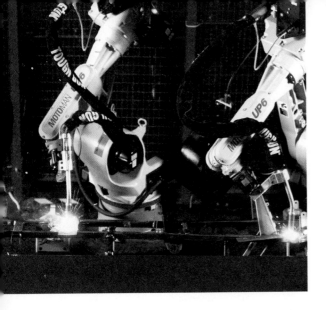

Chapter 9
Computer Systems and Digital Electronics

Outline

9.1 Computer Systems
9.2 Digital Numbering Systems
9.3 Binary Logic Circuits
9.4 Computer Programming

Objectives

Upon completion of this chapter, you will be able to:

- Understand what a bit of information represents.
- Identify the function of a computer system's basic components.
- Explain the basic functions of a computer system.
- Summarize the characteristics and function of the binary number system.
- Apply the conversion formulas for binary, octal, and hexadecimal number systems.
- Identify common types of logic gates.
- Explain how flip-flops are used in binary counters.
- Understand the purpose of instructions in computer programming.

Technical Terms

accumulator
address register
analog information
AND gate
arithmetic logic unit (ALU)
binary-coded-decimal (BCD) number system
binary counter
binary logic circuit
binary number system
binary point
bistable device
bit
bus network
byte
central processing unit (CPU)
control unit
counter
data register
decade counter
decoder unit
digital electronics
digital information
dynamic RAM (DRAM)
electrically erasable programmable read-only memory (EEPROM)
erasable programmable read-only memory (EPROM)
execute

fetch	microprocessor unit	program counter
firmware	(MPU)	programmable read-
flip-flop	MPU cycle	only memory
hexadecimal number	NAND gate	(PROM)
system	NOR gate	random access
input-output (I/O)	NOT gate	memory (RAM)
transfer	octal number system	read-only memory
interrupt	OR gate	(ROM)
inverter	period	register unit
logic circuit	personal computer (PC)	software
logic gate	place value	static RAM (SRAM)
memory	positive logic	truth table

Overview

The electronic circuitry used with robots provides commands that control motors, hydraulic systems, and pneumatic systems. These circuits also store information, count, encode, and decode. The circuits that perform these functions are called *logic circuits*. The technology that controls robotic and other automated systems is called *digital electronics*.

This chapter presents the fundamentals of computer design and function, as they are applied to robots, and discusses the basics of digital electronics.

9.1 Computer Systems

The term "computer" is used to cover a number of functions, but primarily refers to a system that performs automatic computations. Computers range in size and function, from pocket calculators to complex central units that serve an entire organization. Computers process information in two states—digital or analog. *Analog information* varies continuously. An example is temperature, which is in a constant state of change. For example, the mercury indicator of an analog thermometer may read between one degree mark and the next. *Digital information*, by contrast, occurs in separate full units. With a digital clock, for example, the time is always displayed as one unit (second) or the next, but not between units.

Digital computers use two numbers that represent either the presence or absence of voltage. A voltage pulse is usually represented by a one (1). The absence of voltage is indicated by a zero (0). A single pulse is described as a *bit* of information, or a "binary digit." A group of 8 bits produces a *byte*, sometimes called a "binary word."

Computer systems have certain basic parts, **Figure 9-1**, which may be arranged in a variety of different ways. The organization and design of each circuit differs considerably among manufacturers. A basic computer consists of input and output devices, arithmetic logic and control circuitry, and a form of memory.

Figure 9-1. This diagram represents the common components of a personal computer.

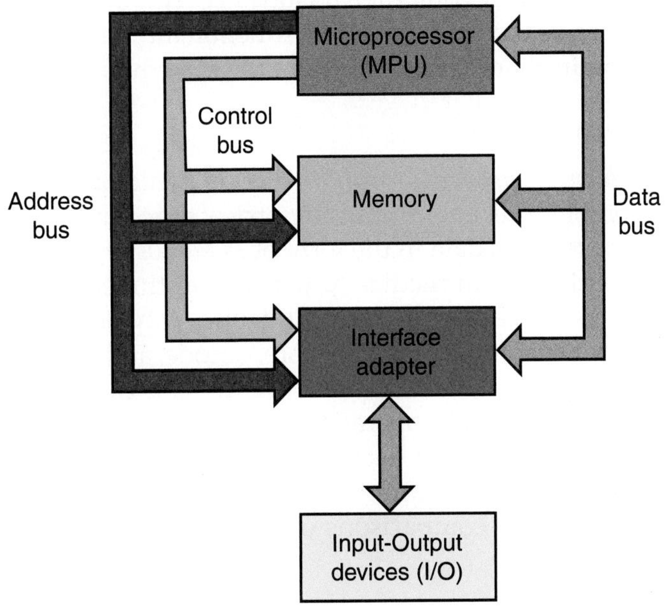

Coded data in the form of 1s and 0s are written into the computer's memory, based on directions provided by a program. This coded data provide operating instructions used to direct the brain of the computer, or the *central processing unit (CPU)*. The CPU "reads" these instructions, processes the data, and carries out programmed operations. Computers typically have several input and output ports that permit the CPU to communicate with external devices, such as monitors, printers, disc drives, jump drives, and modems.

While the basic parts of a computer are the same, the difference among units is in physical size, amount of memory, and processing speed. Mainframes are large computers that have millions of components included in their circuits. Mainframes are used for very large and complex tasks, such as scientific computations and operating industrial plants or governmental departments.

Minicomputer systems are designed for applications that do not require the capacity of a mainframe computer. These computer systems have smaller memory capacity and process data less rapidly than a mainframe system, but are also less expensive. Minicomputers are used for many business, manufacturing, and educational applications.

Personal computers (PCs) are small desktop or portable units built around a single integrated circuit. PCs were called "microcomputers" when first produced in the early 1980s. Today, PCs are widely used in applications ranging from average home use to business management tasks, and architectural drafting to controlling machining cells in factories. While most PCs are standalone units, they may also be wired together to form networks that allow the computers to share information.

Computer System Components

A personal computer includes several components that receive, store, process, and output data. The functionality of each component depends on the performance of all others in the system.

Microprocessor

A *microprocessor unit (MPU)* is a microchip that combines the arithmetic logic unit and control circuitry of the computer, **Figure 9-2**. A microprocessor receives data in the form of 1s and 0s. It may store this data for future processing, or immediately perform arithmetic and logic operations and deliver the results to an output device. A typical microprocessor contains a number of components that store, decode, and process data and commands.

An *arithmetic logic unit (ALU)* is a calculator that performs mathematical and logic operations. It works automatically using signals sent from the instruction decoder. The ALU combines the input from the data register and the accumulator. Its primary operations are addition, subtraction, and logic comparisons using binary numbers.

Figure 9-2. This block diagram of a microprocessor has been simplified for clarity.

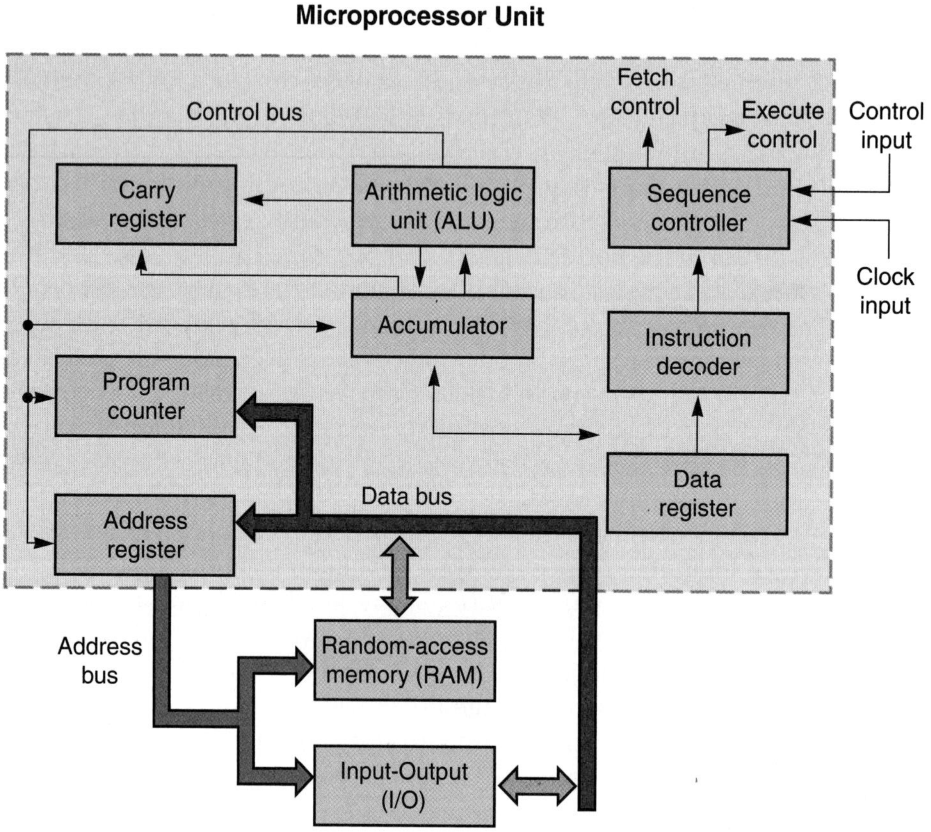

A *decoder unit* examines coded instructions and decides which operation is to be performed by the ALU. After an instruction code is pulled from memory and placed in the data register, it must be decoded. The output of the decoder is sent to the control unit to determine appropriate processing. The *control unit* receives decoded instructions from the decoder and initiates the proper action. The decoded instructions and data are sent to the register unit for appropriate processing.

The *register unit* is a storage component of a microprocessor consisting of several "compartments" that store information. Each compartment in a register unit stores a specific type of data. The *data register* stores information for ALU input. It may also hold an instruction while that instruction is being decoded, or hold data prior to storage in memory. The *address register* temporarily stores the address of a memory location that is to be accessed. The address register is programmable in some units, which means that new instructions may alter its contents. An *accumulator* is a type of register that stores values for mathematical and logic operations in process.

The *program counter* is a memory device that indicates the location in memory of either the instruction currently being performed or the next instruction to be executed. This unit maintains the programmed sequential order of instructions. The sequence may be modified by using subroutines that interrupt the numeric order of instructions and direct the program counter to another instruction or operation.

Buses

The circuits of most microprocessors are connected together by a common bus network. A *bus network* is a series of registers connected together, **Figure 9-3**. An advantage of a bus network is the ease with which data can be transferred. Data is packaged in 8-, 16-, 32-, and 64-bits for transfer between registers. The amount of data a bus can carry is based on the bus width, or capacity.

Information processed by a computer is of two types: instructions and data. A simple addition problem, such as 9 + 2 = 11, can be used to illustrate the distinction. In this problem, the numbers 9 and 2 are data and the addition sign is an instruction. The data is distributed by the data bus to all parts of the system. Instructions (the program) are distributed by the control bus through a separate path. Data may also be removed from memory and distributed as output.

Memory

The use of a microcomputer depends largely on the amount of memory it has available. *Memory* refers to the ability of an MPU to store data, so that a single bit or a group of bits can be easily retrieved. Memory can be added by installing auxiliary chips. The two most common types of memory are ROM and RAM.

Read-only Memory (ROM). Permanent data is stored by a microcomputer's *read-only memory (ROM)*. This information is typically placed in the chip when it is manufactured and often includes operating system and program information. ROM data is permanent and cannot be changed; it is

Figure 9-3. In a computer, registers are connected to buses that serve as paths for data transfer.

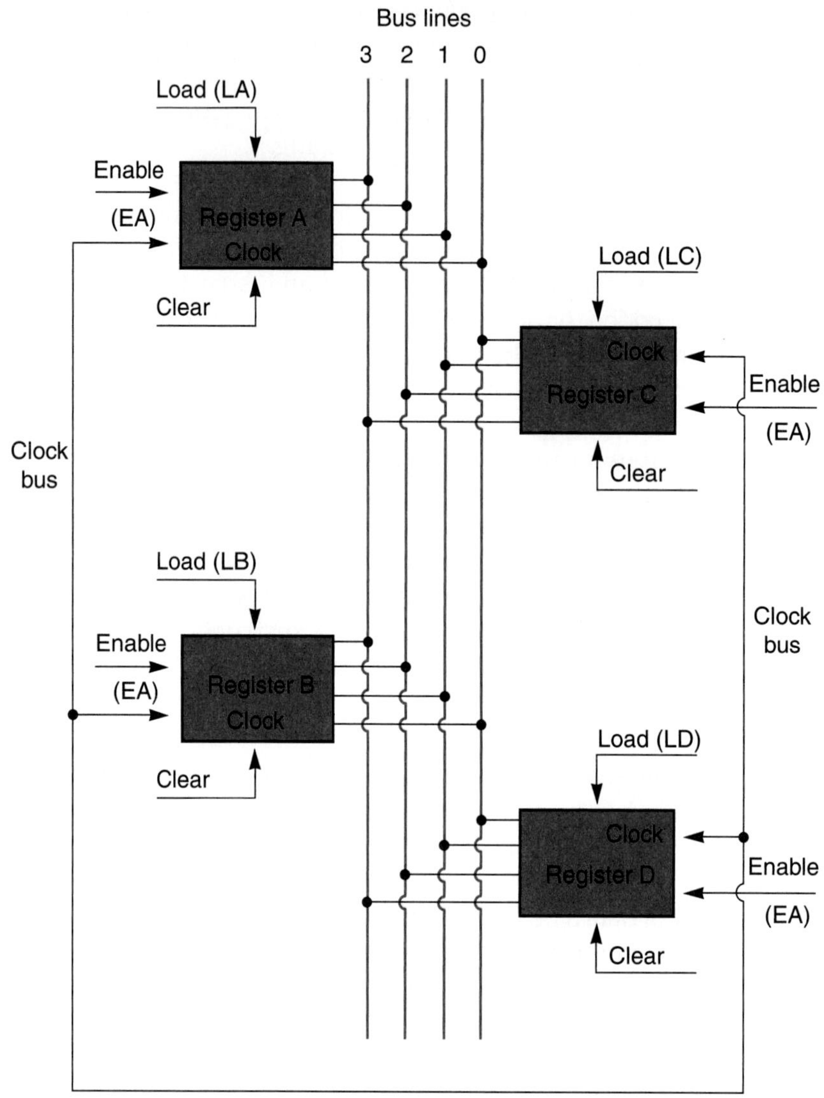

not lost when the MPU power source is turned off. There are several variations of ROM available:

- A *programmable read-only memory (PROM)* chip has data electrically burned onto it and cannot be reprogrammed.
- *Erasable programmable read-only memory (EPROM)* can be erased and reprogrammed. The chip is erased by exposing it to an ultraviolet light.
- *Electrically erasable programmable read-only memory (EEPROM)* can be erased one bit at a time and reprogrammed. Instead of an ultraviolet light, a high electrical charge is used to erase portions of the chip.

Random Access Memory (RAM). Information may be retrieved from *and* written to *random access memory (RAM)*. In other words, the memory can be altered. This type of read/write memory is commonly found in microcomputers because it is very fast. The structure of this chip includes a number of separate circuits, which allow data to be accessed in any order. Data can be placed into memory or retrieved at the same rate. There are two basic categories of RAM available:

- *Dynamic RAM (DRAM)* is a RAM circuit that stores bits of data using separate capacitors. To retain the data, the electrical charge of the capacitors must be constantly refreshed.
- *Static RAM (SRAM)* is a RAM circuit that uses semiconductor devices, called flip-flops, to store data. Flip-flops do not have to be constantly recharged and SRAM transmits data faster than DRAM.

To write data into memory, it is sent along the address bus to a specific address. Data can be read from memory when the microprocessor is turned on. Reading data from memory does not destroy the data. However, a loss of power or turning the unit off will destroy the data. Random access memory is volatile and disappears when power is interrupted.

Basic Functions

Certain functions are basic to most computer systems. Among these are timing, fetch and execute, read memory, write memory, input-output transfer, and interrupt.

Timing

The operation of a computer involves a sequence of instructions. The sequence is controlled by a timing signal. The MPU fetches an instruction, executes the required operations, fetches the next instruction, executes it, and so on, in a cycling pattern. All the actions performed occur at or during a precise time interval. Such an orderly sequence requires a free-running, electronic oscillator or clock. In some systems, the clock may be built into the MPU. In others, timing units feed the system through a separate control bus.

System operations are performed within a timeframe called the *MPU cycle*. For example, the fetch operation takes the same amount of time in each instance for each instruction. An execute operation, however, may consist of many events and sequences. Therefore, execute timing varies a great deal. The time interval for a pulse to pass through a complete MPU cycle, from beginning to end, is called a *period*.

A microprocessor may take a number of clock periods to perform an operation. For example, a chip that can operate at a clock rate of 33 MHz (33 million cycles per second) has a period of 1/33,000,000 or 0.00000003 second for a single cycle. Periods this small are best expressed in microseconds (millionths of a second–μs) or nanoseconds (billionths of a second–ns). At a clock rate of 33 MHz, the length of a single cycle would be 0.3 μs or 30 ns. So, if this chip requires four clock cycles to execute an instruction change, the change would take 0.12 µs or 120 ns. The operating time of an MPU is a good measure of its effectiveness and power.

Fetch and Execute

After programmed information is placed into memory, related action is directed by a series of fetch and execute operations. The sequence is repeated until the entire program has cycled to its conclusion.

The start signal actuates the control section of the MPU, which automatically starts the sequence of operations. The first instruction is to *fetch*, or retrieve, the next instruction from memory. The MPU may then issue a read instruction. The contents of the program counter are sent to memory, which returns the next instruction. The first word of the instruction received is placed into the instruction register. If more than one word is included in an instruction, a longer cycling time results. After the complete instruction is in the MPU, the program counter records one count. The instruction is decoded, and the unit is prepared for the next fetch instruction.

The *execute*, or initiate, operation is based on the instruction that is to be performed. The instruction may be to read memory, write to memory, read the input signal, or transfer to output, among others.

Read Memory

Read memory calls for data to be read from a specific location. The MPU issues a read operation code and sends it to the proper memory address. The read/write memory unit sends the data into the data bus. This number is fed to the MPU, where it is placed in the accumulator after the timing pulse has been initiated.

Write Memory

Write memory calls for data to be stored at a specific location. The MPU issues a write operation code and sends it to a selected read/write memory unit. Data is sent through the data bus and placed into the selected location.

Input-Output Transfer

Input-output (I/O) transfer operations are similar to read/write operations. The major difference is the code number (opcode) used to call up the operation. This code actuates an I/O port, which either receives data from the input or sends it to the output device. For example, data can move in either direction to the read/write memory and flows from the ROM into the data bus, **Figure 9-4**. The output flows from the data bus to the output device.

Interrupt

Interrupt operations are often used to improve efficiency. Interrupt signals come from peripheral equipment—such as keyboards, displays, modems, and printers—to inform the MPU that a peripheral device needs attention. For example, a system is designed to process a large volume of data and output to a printer. The MPU can output data much faster than the printer can produce

Figure 9-4. This simplified view of an MPU address bus illustrates input-output transfer. 16-bit, 32-bit, and 64-bit buses are common in personal computers.

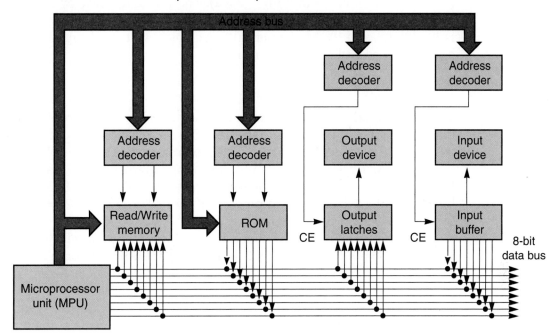

printed pages. This means that the MPU remains idle while waiting for the printer to complete its task. If an interrupt is used, the MPU can output a byte of the data and return to processing functions while the printer operates. When the printer is ready for more data, it sends an interrupt request. The MPU stops, initiates a subroutine to output the next byte, and then continues processing information until it receives another interrupt signal.

9.2 Digital Number Systems

The number system we use in everyday life is the base 10 (decimal) system. In this system, ten digits are used for counting—0, 1, 2, 3, 4, 5, 6, 7, 8, and 9. The number of digits used in a number system determine its base. Because the decimal system uses 10 digits, it has a base of 10.

Number systems assign digits a *place value*. This refers to the position of a digit with respect to the decimal point. In the decimal system, the first position to the left of the decimal point is called the units place. Any digit from 0 to 9 can be used in this place. Number values greater than 9 must be expressed using two or more places. The position to the left of the units place is the 10s place. The number 99 is the largest value that can be expressed using two places in the decimal system. Each place added to the left extends the number value by a power of 10.

A number value in any system can be read by adding the numbers in each place. For example, the decimal value 2583 can be expressed as: $(2 \times 1000) + (5 \times 100) + (8 \times 10) + (3 \times 1)$. The values increase by a power of 10 for each place to the left of the decimal point. In the decimal system, this is 10^3, 10^2, 10^1, and 10^0. Mathematically, each place value is the number times a power of the system base, **Figure 9-5**.

Binary Number System

The decimal number system is difficult to use with electronics, so the binary system is typically used instead. The *binary number system* uses 2 as its base. The largest value that can be expressed by a specific place is the number 1. As a result, only the numbers 0 or 1 are used. The first place to the left of the *binary point* represents units, or 1s. Places to the left of the binary point are expressed in powers of two—$2^0 = 1$, $2^1 = 2$, $2^2 = 4$, $2^3 = 8$, $2^4 = 16$, $2^5 = 32$, $2^6 = 64$, and so on.

When different number systems are used, a subscript number is used to identify the base. The number 100_2 indicates that the binary (base 2) system applies. In the binary system, the number 100 is read as "one-zero-zero" instead of "one hundred."

When converting a binary number to an equivalent decimal number, write down the binary number first. See **Figure 9-6**. Starting at the binary point, indicate the decimal equivalent for each place location where a 1

Figure 9-5. Expressing a base 10 (decimal) number.

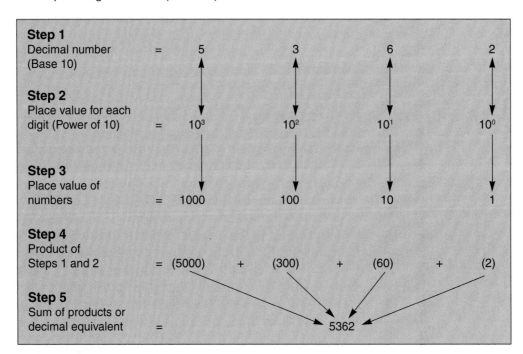

is indicated. For each 0 in the binary number, leave a blank space or indicate a 0. Add the place values and record the decimal equivalent.

Converting a decimal number to its binary equivalent involves repeatedly dividing the decimal number by 2, **Figure 9-7**. When the quotient is a

Figure 9-6. Performing a binary-to-base 10 conversion.

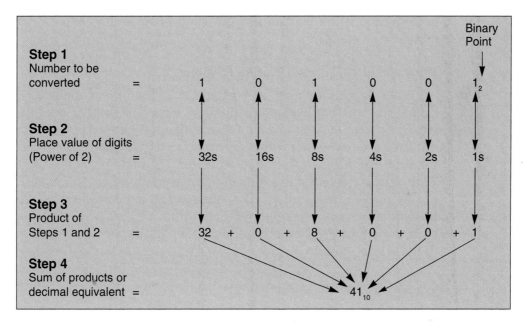

Figure 9-7. A decimal-to-binary conversion is achieved by dividing by 2.

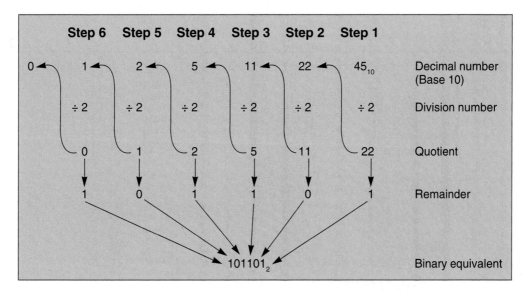

whole number with no remainder, the value is 0. When the quotient has a remainder, the value is 1. Division continues until the quotient is zero.

Electronically, the value of 0 indicates low-voltage or no voltage. The number 1 indicates a voltage larger than 0. Binary systems that use these values are said to have *positive logic*.

The two operational states of a binary system, 1 and 0, are natural circuit conditions. When a circuit is turned off or has no voltage applied, it is in the 0 state. A circuit that has voltage applied is in the 1 state. This makes it possible to change states in less than a microsecond. Therefore, electronic devices can manipulate millions of 0s and 1s per second and process information very quickly.

Binary-Coded-Decimal Number System

When large numbers are written using the binary method, they are difficult to use. For example, $1101001_2 = 105_{10}$. The *binary-coded-decimal (BCD) number system* of counting was devised to overcome this problem.

The base 10 number is first divided into digits according to place value, **Figure 9-8**. For example, the number 105_{10} is made up of the digits 1-0-5. Each digit is converted to a four-digit binary number. In this example, 105_{10} results in $0001\ 0000\ 0101_{BCD}$. The largest digit expressed by any group of BCD numbers is 9. Decimal numbers up to 999_{10} may be converted using only three four-digit binary numbers. Separation between each group of digits is important to include.

Octal Number System

The *octal number system* is a base 8 system used to process large numbers and uses the same principles as the decimal and binary systems. The digits 0, 1, 2, 3, 4, 5, 6, and 7 are used in the place positions. The place values of digits, moving to the left of the octal point, are powers of eight—8^0 = units (1s), 8^1 = 8s, 8^2 = 64s, 8^3 = 512s, 8^4 = 4096s, and so on.

Figure 9-8. When performing a BCD conversion, four binary digits represent each decimal digit.

Given the decimal number		105_{10}	
Step 1 Group the digits	(1)	(0)	(5)
Step 2 Convert each digit to binary group	(0001)	(0000)	(0101)
Step 3 Combine group values	0001 0000 0101 BCD		

The process of converting an octal number to a decimal number is the same as used for binary-to-decimal conversion, but uses powers of 8. See **Figure 9-9**. Converting an octal number to an equivalent binary number is similar to BCD conversion. The octal number is first divided into digits according to place value. Each digit is then converted into an equivalent binary number using only three digits, **Figure 9-10**.

Converting a decimal number to an octal number means dividing by the number 8, **Figure 9-11**. After the quotient has been determined, the remainder is brought down as the place value. When the quotient is even, with no remainder, the value is 0.

Figure 9-9. An octal-to-decimal conversion requires the same basic process as binary-to-decimal conversion.

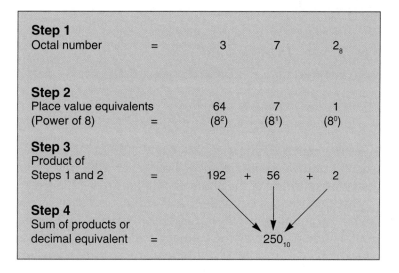

Figure 9-10. In octal-to-binary conversions, a maximum of three binary numbers are used per digit.

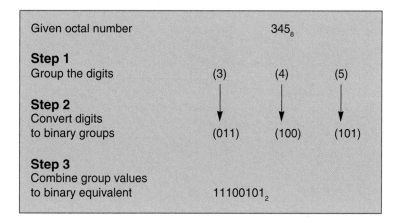

Figure 9-11. A decimal-to-octal conversion involves dividing by 8.

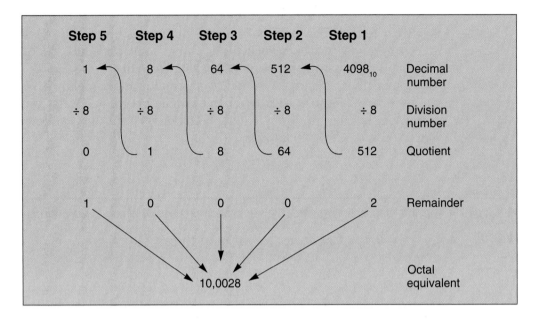

To change a binary number (110100100_2, for example) to an equivalent octal number, the digits must first be divided into groups of three binary numbers starting at the binary point, **Figure 9-12**. Each binary group is then converted into an equivalent octal number. These numbers are combined, while remaining in their respective places, to represent the equivalent octal number. Converting a binary number to an octal number is often done for digital circuits. Binary numbers are first processed at a very high speed.

Figure 9-12. Binary-to-octal conversions are often performed for digital circuits.

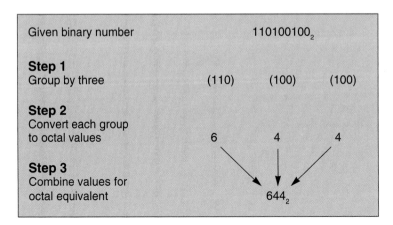

An output circuit accepts the signal and converts it to an octal number to display on a readout device.

Hexadecimal Number System

The *hexadecimal number system* is a base 16 system that is also used to process large numbers. The largest number used in a place is 15, with digits 0–9 and letters A–F. The letters A–F represent digits 10–15, respectively. The place value of digits to the left of the hexadecimal point are powers of sixteen—$16^0 = 1$, $16^1 = 16$, $16^2 = 256$, $16^3 = 4096$, $16^4 = 65,536$, and so on.

The method of changing a hexadecimal number to a decimal number is similar to other conversions, **Figure 9-13**. The hexadecimal number is written in proper digital order and the place values are positioned under each respective digit. The letters are converted to numeric values and each of the values is multiplied by its corresponding place value. The products are added together and the resulting sum is the decimal equivalent of the hexadecimal number.

The process of changing a hexadecimal number to a binary equivalent is a grouping operation, **Figure 9-14**. The hexadecimal number is separated into digits, and each digit is converted to its equivalent in binary form. For example, "D" in hexadecimal notation equals 13 in decimal form, and 1101 in binary. The groups combine to form the equivalent binary number. Note that the leading zeros from the binary number 0010 are dropped.

Decimal-to-hexadecimal conversion is done by dividing by 16, **Figure 9-15**. Remainders can be as large as 15.

Converting a binary number to a hexadecimal equivalent is the reverse of the hexadecimal-to-binary process, **Figure 9-16**. The binary number is

---

Figure 9-13. Hexadecimal-to-decimal conversion involves an extra step—converting the letters to numeric values.

Step 1 Hexadecimal number	=	1	2	C	D_{16}
Step 2 Place values (Power of 16)	=	4096s	256s	16s	1s
Step 3 Convert letters to numbers	=	1	2	12	13
Step 4 Product of Steps 2 and 3	=	4096 +	512 +	192 +	13
Step 5 Sum of products or decimal equivalent	=	4813_{10}			

Figure 9-14. A hexadecimal-to-binary conversion requires grouping the binary numbers.

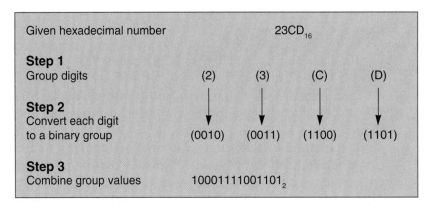

Figure 9-15. A decimal-to-hexadecimal conversion uses a divisor of 16.

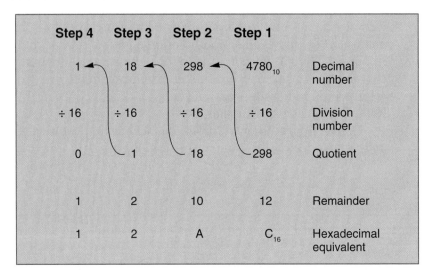

divided in groups of four digits, starting at the binary point. Each group is converted to the equivalent hexadecimal value and the groups are combined to form the hexadecimal value.

9.3 Binary Logic Circuits

Binary signals can be processed easily because they represent two stable states of operation: on or off, 1 or 0, up or down, voltage or no voltage, right or left, and so on. There is no "in-between" state. This makes binary signals far superior to octal, decimal, or hexadecimal signals for use in logic circuits.

Figure 9-16. A binary-to-hexadecimal conversion is performed by creating groups of four binary numbers and converting them to hexadecimal values.

Given binary number				1001101101010_2
Step 1				
Group by four	(0001)	(0011)	(0110)	(1010)
Step 2				
Convert groups to hexadecimal values	1 ↓ 1	3 ↓ 3	6 ↓ 6	10 ↓ A
Step 3				
Combine values for hexadecimal equivalent	$136A_{16}$			

The symbols used to represent an operational state are very important. In positive binary logic, "voltage," "on," or "true" results are indicated by the 1 operational state. "No voltage," "off," or "false" results are indicated by the 0 condition. A circuit can be set to either state and remains at that setting until a change occurs.

Logic Gates

Any electronic device that can be set in one of two operational states—either ON (1) or OFF (0)—by an outside signal is a *bistable device*. These devices include relays, lamps, switches, transistors, and diodes. A bistable device can store one binary digit or bit of information. By using many of these devices, it is possible to build a *binary logic circuit* that makes logical decisions based on input signals. The basic binary logic circuits are the AND circuit, OR circuit, and NOT circuit. These circuits are also called *logic gates*. The term "gate" refers to the circuit's ability to pass or block certain signals. The decision made by each type of circuit is unique. An IF...THEN sentence is often used to describe the basic operation of a logic gate. For example, "IF the inputs applied to a gate are all 1, THEN the output will be 1."

The logic gates discussed in the following sections illustrate basic operation using simple switch and lamp input-outputs. In actual applications, logic gates are typically integrated circuits (ICs).

AND Gates

An *AND gate* has two or more inputs and one output, **Figure 9-17**. When a switch is turned on, the state is 1; when the switch is off, the state is 0. If both inputs are in the 1 state simultaneously, a 1 will be output

Figure 9-17. The AND gate circuit.

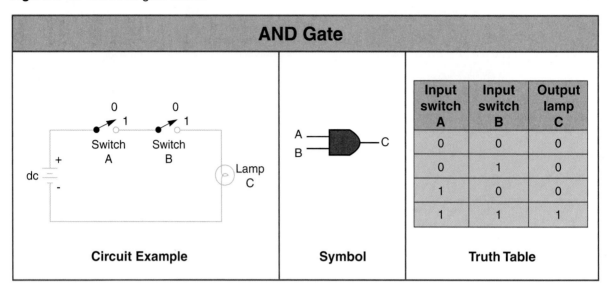

(the lamp will light). A *truth table* shows combinations of inputs and the resulting outputs of a logic gate. The operation of the AND gate is simplified by describing the input-output relationships in the truth table. The AND gate produces a 1 output only when switches A and B are both 1.

Each input to an AND gate can create two operational states: 1 and 0. A two-input AND gate has 2^2, or 4, possible combinations that influence the output. A three-input gate has 2^3, or 8, combinations and a four-input gate has 2^4, or 16, combinations. These combinations are placed in the truth table in binary order. For a two-input gate, for example, the order is 00, 01, 10, and 11. This represents the binary count of 0, 1, 2, and 3 (in order).

OR Gate

An *OR gate* has two or more inputs and one output, **Figure 9-18**. Like the AND gate, each input to the OR gate produces two possible states: 1 and 0. The output is 1 when both switches are 1 *or* when either switch A or B is 1. This gate is used to decide whether or not a 1 appears at either input. The truth table for an OR gate shows that if the state of any input is 1, the output will be 1.

NOT Gate

A *NOT gate* has one input and one output, **Figure 9-19**. The output of a NOT gate is opposite to the input. When the switch is "on" (1 state), it shorts out the lamp. Placing the switch in the "off" condition (0 state) causes the lamp to be "on." A NOT gate is also called an *inverter*.

Figure 9-18. The OR gate circuit.

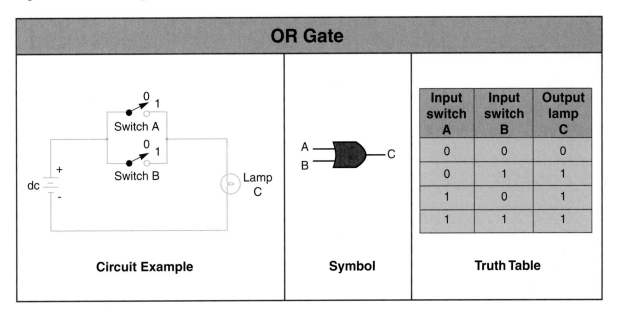

Figure 9-19. The NOT gate circuit.

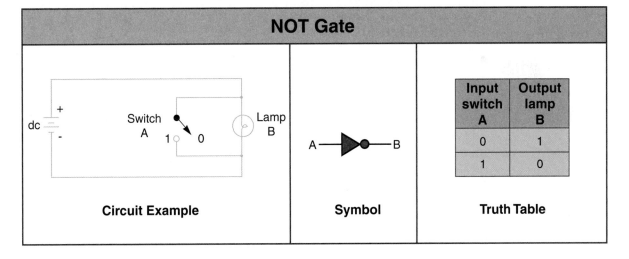

Combination Logic Gates

When a NOT gate is combined with an AND gate, it is called a *NAND gate*. This is an inverted AND gate, **Figure 9-20**. For example, when switches A and B are both "on" (1 state), lamp C is off (0 state). When either or both switches are "off," lamp C is on (1 state).

Figure 9-20. The NAND gate circuit.

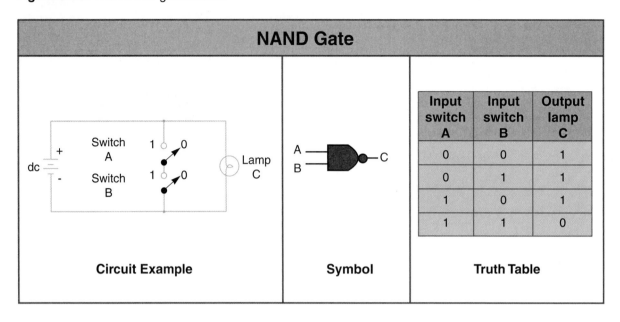

The combination of a NOT gate and an OR gate results in a *NOR gate*, **Figure 9-21**. A NOR gate is the opposite of an OR gate. A 1 state is produced as the output only when A is 0 and B is 0.

Flip-Flops

Flip-flops are memory devices used in digital circuits. They are often the basic logic element for counting, temporary memory, and sequential switching operations. Flip-flops can be used to hold an output state even when the input is completely removed, and can change their output based on an appropriate input signal.

Reset-set (R-S) flip-flops show the different states before an input occurs and how they change afterwards, **Figure 9-22**. Its logic diagram, symbol, and truth table are more complicated than those for a simple logic gate.

Flip-flops operate in step with a clock pulse and must often be set and cleared at specific times with respect to other circuits. Both the appropriate R-S inputs and clock pulse must be present to cause a state change. This device is called an R-S triggered flip-flop, or simply an R-S-T flip-flop, **Figure 9-23**. The truth table for an R-S-T flip-flop is basically the same as that for an R-S flip-flop. A state change occurs only when the clock pulse arrives at the trigger (T) input. A two-input AND gate is added to the set and reset inputs for this purpose.

Figure 9-21. The NOR gate circuit.

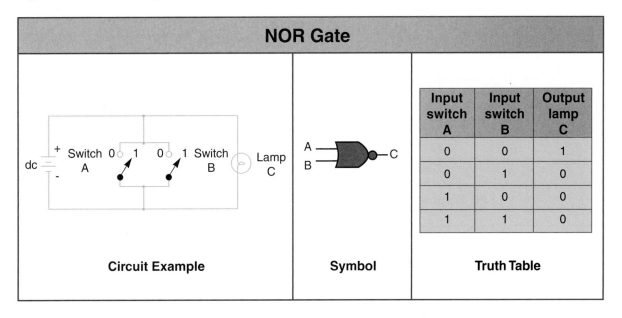

Figure 9-22. The R-S flip-flop.

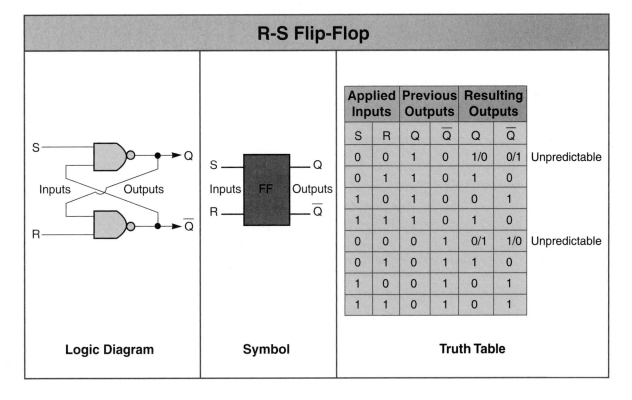

Figure 9-23. The R-S-T flip-flop.

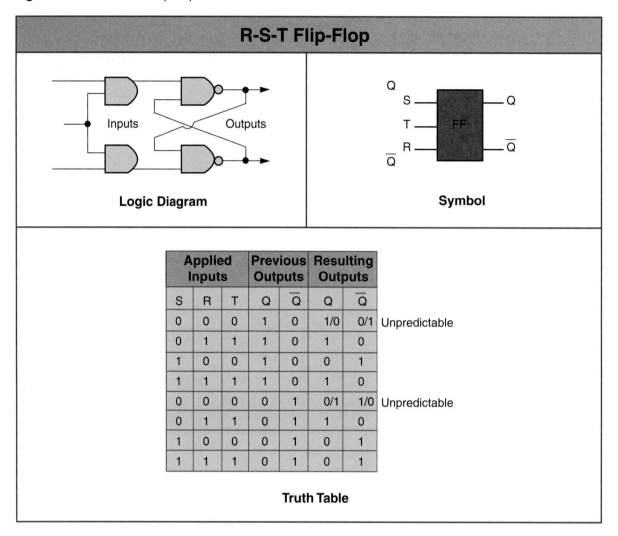

The JK flip-flop is unique, as it has no unpredictable output states (**Figure 9-24**). It can be set by applying a 1 to the J input and cleared by feeding a 1 to the K input. A 1 signal applied to both J and K inputs simultaneously causes the output to change states, or toggle. A 0 applied simultaneously to both inputs does not initiate a state change. The inputs of a JK flip-flop are controlled directly by clock pulses. Several variations on the basic JK flip-flop are available, including preset and preclear inputs that are used to establish sequential operations at precise times.

Digital Counters

One of the most versatile and important logic devices is the *counter*. Counters are used to count a wide variety of objects in various applications.

Figure 9-24. The JK flip-flop.

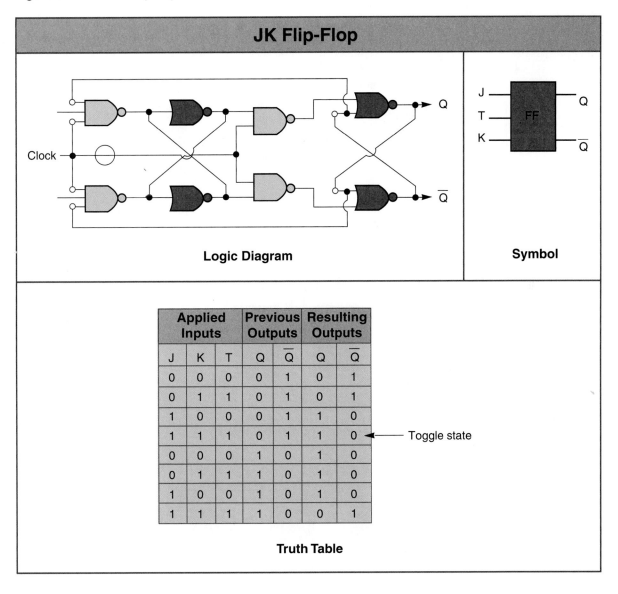

However, they really count only one thing—electronic pulses. These pulses may be produced mechanically, acoustically, with a clock mechanism, or by a number of other processes. Two common forms of this device are binary counters and decade (BCD) counters.

Binary Counters

In *binary counters*, flip-flops are connected so that the output of the first circuit (Q) drives the trigger, or clock input, of the next circuit. See **Figure 9-25**. Each flip-flop, therefore, has a divide-by-two function.

Figure 9-25. Some digital counters that use a JK flip-flop.

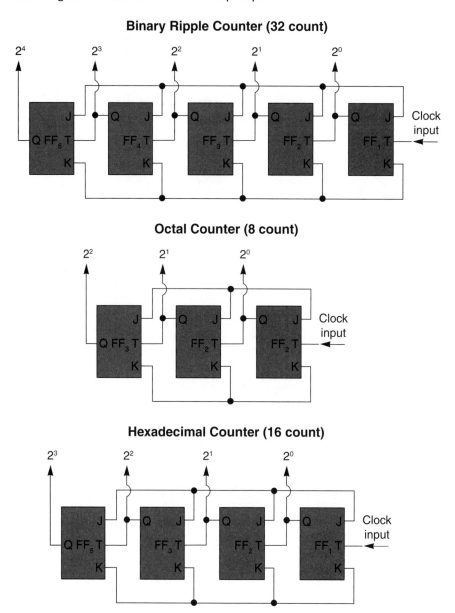

In a binary ripple counter, the J and K inputs for each flip-flop are held at a logic 1 level. Each clock pulse applied to the input of the first flip-flop causes a change in state. Since the flip-flops trigger only on the negative-going part of the clock pulse, the output of FF_1 alternates between 1 and 0 with each pulse. Thus, a 1 output appears at Q of FF_1 for every two input pulses. This means that each flip-flop has a divide-by-two function. Five flip-flops connected in this manner produce a 2^5, or 32 count. The largest count in this case is 11111_2 (31_{10}). The next applied pulse clears the counter so that 0 appears at all the Q outputs.

Grouping three flip-flops together creates a binary-coded octal (BCO) counter. For example, the number 111_2 represents the seven count, or seven units, of an octal counter. Two groups of three flip-flops connected in this manner produce a maximum count of $111,111_2$, which represents 77_8, or 63_{10}.

It is possible to develop the units part of a binary-coded hexadecimal (BCH) counter by placing four flip-flops together in a group. Thus, 1111_2 represents F_{16} or 15_{10}. Two groups of four flip-flops could produce a maximum count of $1111,1111_2$, which represents FF_{16} or 255_{10}. Each succeeding group of four flip-flops raise the count to the next power of 16.

Decade Counters

Since most of the mathematics that we use today is based upon the decimal (base 10) system, it is important to be able to count by this method. The output of a binary counter must be changed into decimal form before it can be used in this way. The first step is to change binary signals into binary-coded decimal (BCD) form using a *decade counter*. For example, a four-bit binary counter achieves 16 counts using four flip-flops, **Figure 9-26**. To convert this counter into a decade counter, it must skip some of its counts, **Figure 9-27**. In this example, the first seven counts occur naturally. Therefore, FF_D remains at 1. This is applied to the J input of FF_B, permitting it to trigger with each clock pulse. At the seventh count, 1s appearing at the Q outputs of FF_B and FF_C are applied to the AND gate. This action produces a logic 1 and applies it to the J input of FF_D. The next clock pulse triggers FF_A, FF_B, and FF_C into the off state and turns on FF_D. This represents the eighth count.

When FF_D is in the "on" state, Q is 1 and $\overline{Q}$ is 0. This causes a 0 to be fed to the J input of FF_B, which prevents it from triggering until cleared. The next clock pulse causes FF_A to be set to a 1. This registers a 1001_2, which is the ninth count. The next count clears FF_A and FF_D instantly. Since FF_B

Figure 9-26. A 4-bit binary counter uses four flip-flops. The counts are listed below the binary counter, reading right to left.

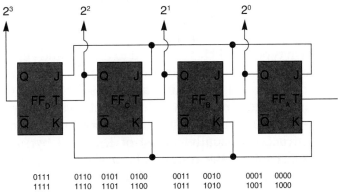

Figure 9-27. Converting a 4-bit binary counter into a decade (BCD) counter.

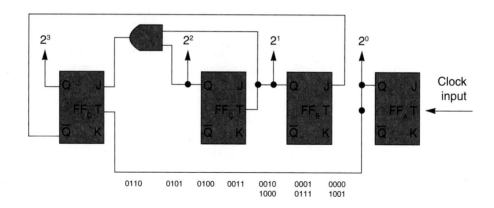

and FF_C were previously cleared by the seventh count, all 0s appear at the outputs. The counter has, therefore, cycled through the ninth count and returned to zero, ready for the next input.

9.4 Computer Programming

In microcomputer systems, some programs are held in read-only memories called *firmware*, and others are stored in the form of *software*. Firmware is placed on ROM chips, which requires that program changes be made by changing the ROM. Software has the greatest flexibility. It is transferred to the system by keyboard, CD-ROM, or download, and the instructions are stored in read/write memory.

Instructions

Instructions normally appear as a set of characters or symbols that define a specific operation. These symbols are similar to those on a computer keyboard, and include decimal digits 0–9, letters A–Z, and, in some cases, punctuation marks and special characters. Instructions may also appear in the form of binary numbers, hexadecimal numbers, or mnemonic codes. See **Figure 9-28**. Each computer system has a number of instructions of this 1-byte type that contain only an opcode.

Each type of MPU is designed to understand certain set of instructions. These instructions are in the form of binary data, normally held in a read-only memory unit. The unit is address-selected and connected to the MPU through the common data bus. Instructional sets are an example of firmware—they are fixed at a specific location and cannot be changed.

Instructions usually consist of 1, 2, or 3 bytes of data. This type of data must follow the commands in successive memory locations. These commands are usually called addressing modes.

Figure 9-28. Example of opcode instructions. Note that the code is in hexadecimal form, and a mnemonic is given for each one.

Mnemonic	Opcode	Meaning
ABA	1B	Add the contents of accumulators A and B. The result is stored in accumulator A. The contents of B are not altered.
CLA	4F	Clear accumulator A to all zeros.
CLB	5F	Clear accumulator B.
CBA	11	Compare accumulators: Subtract the contents of ACCB from ACCA. The ALU is involved, but the contents of the accumulators are not altered. The comparison is reflected in the condition register.
COMA	43	Find the ones complement of the data in accumulator A, and replace its contents with its ones complement. (The ones complement is simple inversion of all bits.)
COMB	53	Replace the contents of ACCB with its ones complement.
DAA	19	Adjust the two hexadecimal digits in accumulator A to valid BCD digits. Set the carry bit in the condition register when appropriate. The correction is accomplished by adding 06, 60, or 66 to the contents of ACCA.
DECA	4A	Decrement accumulator A. Subtract 1 from the contents of accumulator A. Store the result in ACCA.
DECB	5A	Decrement accumulator B. Store in accumulator B.
LSRA	44	Logic shift right, accumulator A or B. $0 \rightarrow b_7\ b_6\ b_5\ b_4\ b_3\ b_2\ b_1\ b_0 \rightarrow C$
SBA	10	Subtract the contents of accumulator B from the contents of accumulator A. Store results in accumulator A.
TAB	16	Transfer the contents of ACCA to accumulator B. The contents of register A are unchanged.
TBA	17	Transfer the contents of ACCB to accumulator A. The contents of ACCA are unchanged.
NEGA	40	Replace the contents of ACCA with its twos complements. This operation generates a negative number.
NEGB	50	Replace the contents of ACCB with its twos complements. This operation generates a negative number.
INCA	4C	Increment accumulator A. Add 1 to the contents of ACCA and store in ACCA.
INCB	5C	Increment accumulator B. Store contents in ACCB.
ROLA	49	Rotate left, accumulator A.

One-byte instructions are called inherent-mode instructions and are designed to send data to the accumulator registers of the ALU. No address code is needed because it is an implied machine instruction. For example, the instruction "CLA" is a 1-byte opcode that clears the contents of accumulator A. No address or specific definition of data is needed. Inherent mode instructions differ a great deal among manufacturers.

Immediate Addressing

Immediate addressing is done with a 2-byte instruction that contains an opcode and an operand. The opcode appears in the first byte, followed by the 8-byte operand. A common practice is to place intermediate addressing

Robotics in Society: RQ-4 Global Hawk

The RQ-4 Global Hawk is an unmanned aerial vehicle (UAV) built by the Northrop Grumman Corporation and used by the United States military forces. This aircraft has a wingspan of about 116 feet, stands about 15 feet tall, and can travel at a maximum altitude of 65,000 feet (about 19 km)—just on the edge of the stratosphere—without a human pilot onboard.

The Global Hawk takes off, performs its mission, and lands using the programmed data and monitoring provided by two land-based control systems: the Mission Control Element (MCE) and the Launch and Recovery Element (LRE). Pilots operating the MCE manage the sensors, cameras, and indicators onboard, as well as the aircraft itself. The LRE team on the ground is responsible for loading the mission plan into the aircraft and monitoring performance during takeoff and landing.

The Global Hawk is equipped with wideband satellite links, line of sight links, and a 48" satellite antenna for communication and data transmission with the land-based control systems. All systems and sensors onboard gather and transmit data and images day or night, regardless of weather conditions. Having this real-time information has proven critical in military combat operations. The Global Hawk is improving the efficiency, accuracy, and strategy of military operations, which affects the safety of troops fighting on the ground.

Effective use of the Global Hawk requires knowledge of its complex communication and processing systems. (Courtesy of Northrop Grumman Corporation)

instructions in the first 256 memory locations. Since this is the fastest mode of operation, these instructions can be retrieved very quickly.

Relative Addressing

Relative addressing instructions transfer program control to a location other than the next consecutive memory address. Transfer is often limited to a number of locations in front or in back of the present location. The 2-byte instruction contains an opcode in the first byte and a memory location in the second byte. The second byte points to the location of the next instruction which is to be executed.

Indexed Addressing

Indexed addressing is similar to relative addressing. The second byte of the 2-byte instruction is added to the contents of the index register to form a new, or "effective," address. This address is obtained during execution, rather than being held at a predetermined location. It is held in a temporary memory address register to ensure that it does not get altered or destroyed during processing.

Direct Addressing

In direct addressing, the address is located in the byte of memory following the opcode. This permits addressing the first 256 bytes of memory, from 0000_{16} to $00FF_{16}$.

Extended Addressing

Extended addressing increases the ability of direct addressing to accommodate more data. It is used for memory locations above $00FF_{16}$ and requires 3 bytes of data for the instruction. The first byte is a standard opcode. The second byte is an address location for the most significant 8 bits of data. The third byte holds the address of the least significant 8 bits of data being processed.

Program Planning

A computer cannot solve even the simplest problem without the help of a well-defined program. The system follows this program to accomplish a task. This is why programming is an essential part of all computer applications. The programmer must be fully aware of the instructions used by the system.

A programmer should be able to decide what specific instructions are needed to perform a given task. A limited number of operations can usually be developed without the aid of a diagrammed plan of procedure. Complex problems, however, require a specific plan in order to avoid confusion or the loss of an important step. Flowcharts are commonly used to aid in this type of planning, **Figure 9-29.**

Figure 9-29. These symbols are used by programmers when constructing flowcharts.

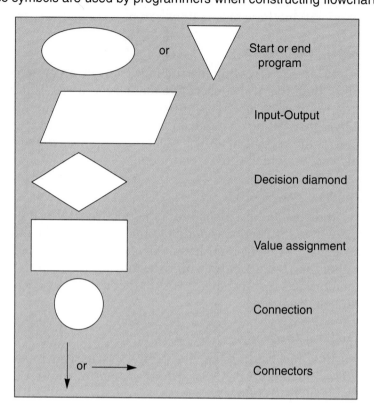

Review Questions

Write your answers on a separate sheet of paper. Do not write in this book.

1. Explain the difference between *analog information* and *digital information*.
2. What is a *bit* of information? How many bits in a *byte*?
3. What is the function of the *CPU*?
4. Define *register unit* and list examples of different types of registers.
5. What is the difference between *ROM* and *RAM*?
6. List and define the functions that are basic to most computer systems.
7. Convert the binary number 11001 to its decimal equivalent.
8. What is the base factor for the hexadecimal number system?
9. What is a *bistable device*?
10. How many inputs and outputs can be found on an OR gate?
11. What is an *inverter*?
12. How are flip-flops used in binary counters?
13. Explain the difference between *firmware* and *software*.

Learning Extensions

1. Research the history and applications of the binary number system. Prior to computer programming, what were some historical applications of the binary system?

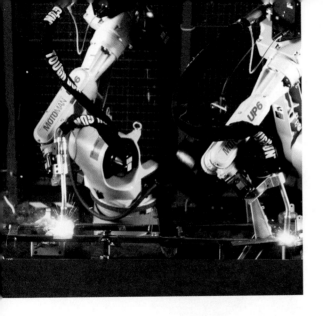

Chapter 10
Interfacing and Vision Systems

Outline

10.1 Interfacing
10.2 Machine Vision

Objectives

Upon completion of this chapter, you will be able to:
- Understand the purpose of interfacing robotic systems.
- Explain the hardware options for interfacing multiple systems.
- Identify the types of robotic operations that use machine vision systems.
- Explain the four functions that take place during image processing.

Technical Terms

digital input port
digital output port
general-purpose interface bus (GPIB)
I/O port
image acquisition
image analysis
image interpretation
image preprocessing
interface
machine vision system
parallel port
parallel transmission
pixel
serial port
serial transmission

Overview

Robots do not work in isolation. They interact continuously with other production equipment in a process called *interfacing*, **Figure 10-1**.

Vision systems give a robot "eyes" and allow it to perform a wider variety of tasks. Machine vision systems are built into automated work cells and are used to determine part orientation, handle measurement, and inspect parts, among other tasks.

This chapter discusses the principles and components of interfacing, and covers the functionality and applications of machine vision systems.

241

Figure 10-1. In a robotic work cell, robots interface with other equipment. (Motoman)

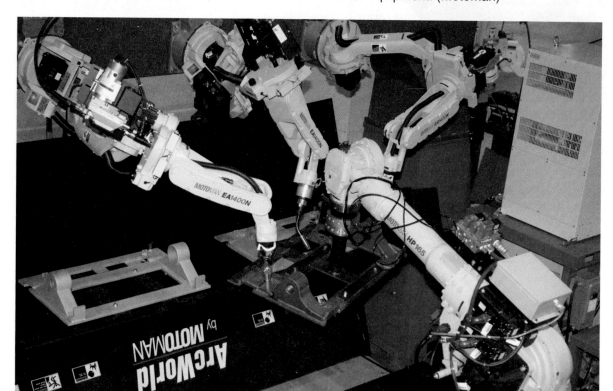

10.1 Interfacing

An *interface* is the common point at which two or more systems communicate with each other. External end effector sensors, work cell limit switches, external relays, operator alarms, safety equipment, bar code readers, conveyors, transfer lines, programmable logic controllers, and machine vision systems all interface with robots.

A robot must be equipped with *I/O ports* (input/output) to communicate with other systems. The digital signals used for communication travel from system to system through these ports. Common I/O ports include digital input ports, digital output ports, serial ports, and parallel ports.

Digital Input Ports

Robots receive a signal indicating the state of an external switch (on or off) through a *digital input port*. Limit switches, **Figure 10-2**, communicate information about such things as the presence or absence of a part at a given location. This information is used by the robot's computer program

Figure 10-2. Limit switches communicate information through input ports. This limit switch is used to limit a robot's rotary motion. (Motoman)

to initiate some action, **Figure 10-3**. A common use of digital input ports is with safety devices. Safety devices, such as light curtains, motion detectors, and door switches on safety fences, may be connected to a digital input port, which sends a signal to the controller to shut down the robot when a person or object enters the work envelope. Many robots are equipped with 16 or more digital input ports.

Digital Output Ports

One way that a robot interacts with external equipment, such as a conveyor belt drive motor, solenoid, or another robot, is through *digital output ports*. Digital output ports permit digital signals to be sent from a robot controller to equipment controls, **Figure 10-4**.

Whenever a control signal from the robot's computer program is at a low voltage level, the base (control element) of transistor Q1 does not conduct. The output current flows from the emitter (current source) to the collector (output element). When the base is at a low voltage level, the output current is zero and the circuit is high resistance. The motor control relay (CR) is not energized and the motor is OFF.

When the computer sends a high-level signal (usually 12V or 24V) to the base control element of transistor Q1, the current flow from the emitter to the collector is at maximum. The resistance between the emitter and collector is now very low. In this condition, the control relay is energized and the motor is turned ON. Robot controllers are equipped with several digital output ports and a matching number of input ports.

Figure 10-3. This illustration shows a limit switch connected to a typical digital input port. When no object is present, the limit switch is open. When an object, such as an assembly part, is present, the limit switch closes and conducts 24V direct current to the transistor. The controller circuitry detects this change in condition (presence of an object and a change of transistor state), and passes the information to the robot's computer program to determine the appropriate action.

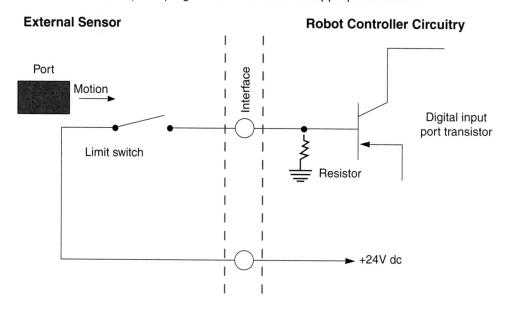

Figure 10-4. When enough voltage is applied to digital output port transistor Q1, the lower end of the control relay is grounded. Grounding the lower end of the control relay coil causes the relay contacts to close. This supplies ac voltage to the conveyor belt motor.

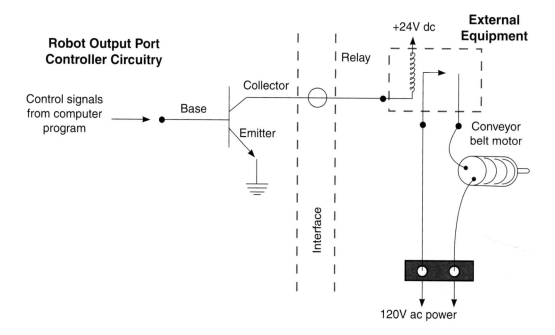

Serial Ports

Serial transmission delivers digital data over long distances, but at a fairly slow rate. The transmission rate is slow because data must be transmitted one bit at a time. Two computers can communicate with each other over telephone circuits via *serial ports* connected to modems. Existing transmission lines can be used, which is an advantage.

A common serial interface used in robotics is the standard RS-232C. RS-232C was developed to standardize the interface between data terminal equipment and data communications equipment. This interface is most commonly used when binary data is transmitted over short distances, such as between two computers or between a computer and a piece of peripheral equipment. The circuit performs three functions: data transfer, timing, and control (**Figure 10-5**).

Figure 10-5. This chart lists pin assignments and interchange circuit functions for the RS-232C serial interface standard.

Pin Assignments and Interchange Circuit Functions

Interchange Circuit	Pin Number	Description	Ground	Data		Control		Timing	
				From DCE	To DCE	From DCE	To DCE	From DCE	To DCE
AA	1	Protective ground	X						
AB	7	Signal ground/common return	X						
BA	2	Transmitted data			X				
BB	3	Received data		X					
CA	4	Request to send					X		
CB	5	Clear to send				X			
CC	6	Data set ready				X			
CD	20	Data terminal ready					X		
CE	22	Ring indicator				X			
CF	8	Received line signal detector				X			
CG	21	Signal quality detector				X			
CH	23	Data signal rate selector (DTE)					X		
CI	23	Data signal rate selector (DCE)				X			
DA	24	Transmitter signal element timing (DTE)							X
DB	15	Transmitter signal element timing (DCE)						X	
DD	17	Receiver signal element timing (DCE)						X	
SBA	14	Secondary transmitted data			X				
SBB	16	Secondary received data		X					
SCA	19	Secondary request to send					X		
SCB	13	Secondary clear to send				X			
SCF	12	Secondary received line signal detector				X			

Applications that do not involve long-distance communications require only a few RS-232C connections. These are the signal ground (AB), transmitted data (BA), and received data (BB), **Figure 10-6**. Long distance communications (those involving modems) require many other connections.

Another common use of the RS-232C serial port is to connect a robot to a PC. A PC can be used offline to write and debug the initial program. The program can then be downloaded to the robot, using the RS-232C port, to provide the robot with the exact moves needed for an application.

When preparing cables to be used with the RS-232C port, do not exceed a total cable capacitance of 2500 picofarads (pF). The capacitance limits the distance between the two pieces of equipment. Another limitation is a 20-kilobit maximum data transfer rate. Although the RS-232C standard does not specify the data format, such as 8 bit ASCII.

Parallel Ports

Parallel transmission sends multiple bits of data at the same time, following side-by-side paths, like the lanes of a highway. This method allows digital equipment to communicate with peripheral equipment using *parallel ports*. Although parallel transmission is faster, it is generally considered too costly to use over long distances.

One commonly used parallel port standard is the IEEE 488. It was developed in 1975 by the Institute of Electrical and Electronic Engineers to reduce the amount of time needed to set up test equipment. At that time, test equipment had its own independent interface and interfacing more than two devices to a computer could be complicated and expensive. The IEEE 488 standard allows computer-controlled test equipment to be set up in a matter of hours, which benefits any industry using automated testing.

The IEEE 488 standard applies when a computer is connected to measuring equipment using a *general-purpose interface bus (GPIB)*. The GPIB is a cable that interfaces system controllers with programmable instruments, and can support a maximum of 15 devices, **Figure 10-7**. Total cable length is 20 meters, or 2 meters times the

Figure 10-6. The external video terminal is interfaced to a robot controller's microprocessor module via RS-232C connections. This illustration shows how AB, BA, and BB connections communicate between the microprocessor and the video terminal.

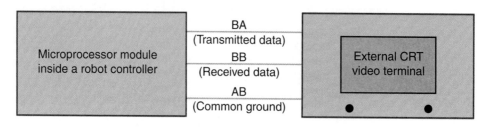

Figure 10-7. The GPIB parallel interface bus is used here to connect external test equipment to a digital system controller or PC.

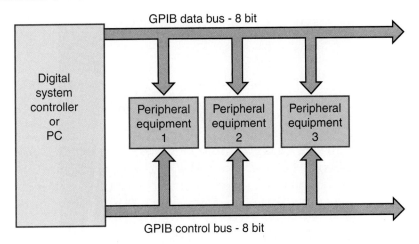

number of connected devices, whichever is less. The cable between any two devices must not exceed 4 meters. The bus carries data in both directions, 8 bits at a time, and controls both the data and the direction. The bus can connect various kinds of devices to the controller, such as equipment that can only "talk," equipment that can only "listen," and equipment that can both "talk" and "listen." Many controllers can be attached to the GPIB, but only one controller may have command at any given time.

10.2 Machine Vision

Machine vision systems give a robot "eyes," which allow it to perform in a more intelligent manner. These systems may be used for such applications as guidance, part orientation, measuring and inspecting parts, and image identification, **Figure 10-8**. Machine vision has become an essential requirement in automated work cells. Robotic work cells that are not equipped with machine vision require all parts to be pre-oriented before a robot can grasp them and perform its task.

Machine vision eliminates the need for many machine operators by performing many of the complex tasks once done only by humans. Systems used for assembly and inspection have had a major impact in the automotive and electronics industries. For example, today's printed circuit boards often contain thousands of individual parts. If even a tiny fraction of these parts is incorrectly placed on the board, the cost of finding and repairing the problem is enormous. When human workers perform circuit board assembly, fatigue can cause as much as a 20 percent error rate. Using robots with machine vision for circuit board assembly significantly reduces the error rate.

Figure 10-8. The video camera "eyes" of this system are suspended over the conveyor to scan parts as they travel along the conveyor. (FANUC Robotics)

Fundamentals of Machine Vision

Machine vision systems use video cameras and computers to translate light energy into an image. These images are then used to determine part orientation or another designated task. However, even the most effective machine vision systems cannot achieve the level of perception and sophistication of the human eye.

Some of the earliest machine vision systems were introduced in the late 1970s. Early systems were expensive due to the high cost of the computer memory needed to handle the images. They were also inefficient, because all computers, except mainframes, contained fairly slow processors. Today, however, faster computers and expanded memory capabilities make vision systems very practical. The data obtained can be used to make decisions on the processing steps or other action taken by either human operators or a robot. Vision systems can also be used in conjunction with material inventory and material flow within a factory.

Four functions take place during image processing—acquisition, preprocessing, analysis, and interpretation, **Figure 10-9.**

Image Acquisition

Image acquisition involves illuminating a workpiece and digitally scanning its image. Fluorescent lamps, incandescent bulbs, strobe lights, or arc lamps usually provide the illumination. The type and amount of light needed depends on the application. Front lighting is used to enhance surface features, such as bar codes or labels. Side lighting is used when three-dimensional (3D) images are desired. Back lighting provides a silhouette of the object.

Scanning an image is done with a video camera. The solid-state digital cameras used with vision systems have either a charge-coupled device (CCD) or a charge-injection device (CID). At the heart of the camera is a silicon chip used with an array of photosensitive elements. The light reflected into the camera lens from the workpiece falls onto this photosensitive array and is converted into an analog electrical signal. The image is broken down into dots of light called picture elements, or *pixels*. Pixels are the means used to form the image on a TV screen. A typical digital video camera can produce an image that measures 1024 by 768 pixels.

Image Preprocessing

During *image preprocessing*, an analog-to-digital (AD) converter changes the analog signal into an equivalent digital signal, **Figure 10-10**. The digital signal represents light intensity values over the entire image. These values are stored in the memory, which allows the digital image to be analyzed and interpreted.

Figure 10-9. These digital electronic subsystems are found in typical machine vision systems.

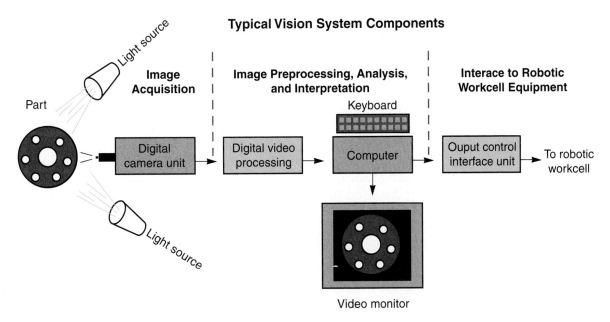

Figure 10-10. This vision system provides 100 percent inspection of the manufactured pieces. Any defects found in the pieces are automatically reported by type, location, and size on a color display screen. (Adept Technology, Inc.)

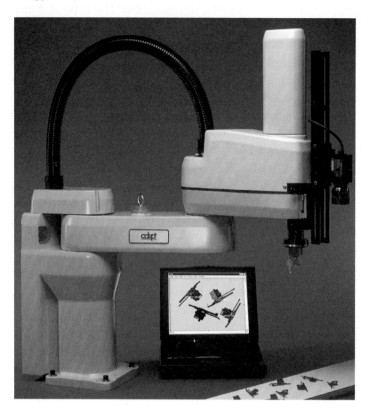

Image Analysis and Interpretation

During *image analysis*, information from the image is gathered and analyzed by computer software. Using algorithms, the software identifies and measures features of the digital image. After all the features are analyzed, the image information is interpreted. The goal of a machine vision system is *image interpretation*, which enables a robot to make decisions about the tasks that must be performed.

Image analysis computer programs, **Figure 10-11**, provide a powerful computer-based development environment with flexible programming, debugging, and testing capabilities. Vision guidance and inspection applications may be used to enhance a variety of robotic operations. Vision-guided robots are used to improve the accuracy of machining, handling, and other industrial operations. Image analysis for inspection is commonly used to collect and record dimensional measurements. This provides a nondestructive method of quality control testing.

Figure 10-11. Image analysis is a common method used in quality control processes. A—Measurements can be specified on a live preview image with tolerances specified. B—The inspection of other parts can be based on the specifications established for the live preview image. Inspection can immediately determine if measurements fall within the specified tolerance.

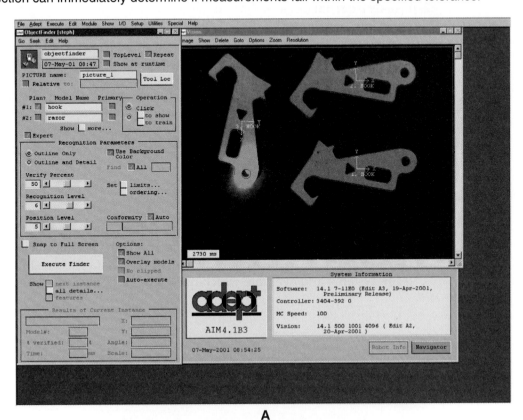

A

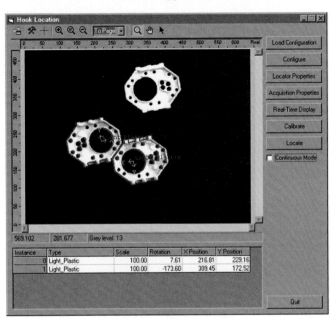

B

Review Questions

Write your answers on a separate sheet of paper. Do not write in this book.

1. What is an *interface* and how does it apply to robotics?
2. List six external devices that can be connected to a robot's digital input and output ports.
3. To send digital data from a robot to external equipment at the highest possible transmission rate, would you use a parallel or a serial port? Why?
4. Explain the advantages and disadvantages of serial ports.
5. What is a general-purpose interface bus (GPIB)?
6. Identify three operations that may use a robotic vision system.
7. Explain the four functions or phases that take place during image processing.

Learning Extensions

1. Conduct a Web search to identify vision system manufactures. Review the system specifications and capabilities of three different vision system products. Create a chart to compare the data gathered for all three systems and evaluate which qualities make each system unique.

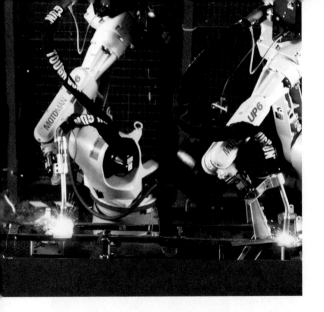

Chapter 11
Maintaining Robotic Systems

Outline

11.1 Troubleshooting
11.2 General Servicing Techniques
11.3 Preventive Maintenance

Objectives

Upon completion of this chapter, you will be able to:
- Establish a workable troubleshooting plan.
- Follow the proper steps to identify equipment problems.
- Develop and implement a preventive maintenance plan.

Technical Terms

preventive maintenance
resin-core solder
rework
scrap
solder sucker
troubleshooting

Overview

Robotic systems are complex because they involve many different areas of technology, **Figure 11-1**. These areas include hydraulic, pneumatic, and electrical power, electronic control, and digital computers. Incorporating a variety of technologies makes the maintenance of robotic systems complex.

In this chapter, robotic system maintenance is discussed. Emphasis is placed on an organized approach to troubleshooting, general servicing techniques, and developing and implementing a preventive maintenance program.

Figure 11-1. This operation uses two robotic systems used for welding. All axes of motion are controlled by electric servo motors. Notice that a power supply and associated electronic control equipment are in the area behind the robots. This is a typical industrial application. Other systems use hydraulic or pneumatic power, rather than electrical. (Motoman)

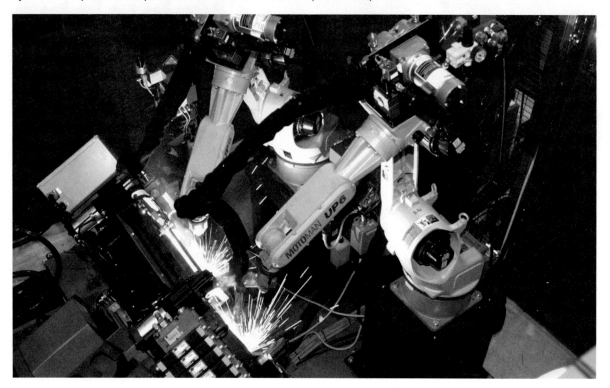

11.1 Troubleshooting

Troubleshooting is the systematic process of identifying and correcting problems in an operation or system. Some problems are easily solved and require little time. Others are more difficult—they occur intermittently or are complex and require many hours of concentration and work. Characteristics of successful troubleshooting include:

- Using common sense.
- Following a logical sequence.
- Knowledge of robotic systems and their operation.
- Skill in using test equipment.
- Ability to read and use schematic diagrams effectively.

To begin troubleshooting any type of system or process, organize your thoughts and identify possible courses of action. Without a well-organized approach, troubleshooting can become a time-consuming guessing game. Use a planned approach to save time and resources.

The initial inspection is an important step in troubleshooting. When this inspection is performed properly, many problems can be efficiently diagnosed. Look for obvious signs—use your senses of sight, touch, smell, and hearing. If a specific part is suspected, turn off the equipment and examine it carefully. A thorough visual inspection should be completed before actual circuit or system testing. Look for:

- Burned parts. They may be charred, blistered, or discolored, and may even have holes.
- Broken parts. Breaks may appear in the form of cracks, wires pulled out of parts, or parts that have been completely destroyed.
- Broken wires and poor connections. These can be located visually and often cause the system to operate inefficiently.
- Smoke or heat damage. A part that smokes when the equipment is turned on indicates a damaged part, but does not necessarily identify the cause of the damage.
- Oil, air, or water leaks. These can be located visually and should be corrected quickly.
- Loose, damaged, or worn parts. These can be located visually or by touch.
- Noisy parts, such as motor bearings. Uncommon noises often indicate defective parts.

Keep in mind that most problems are component failures. Knowing what each component is supposed to do helps in determining the troubles it can cause. Properly test each component. To avoid duplicating effort, make a list of circuit or system operations you have already tested.

Careers in Robotics: Field Service Technician

Field service technicians install, repair, maintain, and provide general support for robotic systems at customer sites. Emergency troubleshooting onsite and over the phone requires expert familiarity with the robotic system and its applications. To ensure the smooth and appropriate operation of robotic systems installed at customer sites, field service technicians provide training for the day-to-day operators and technicians.

To enter this field, typically an associate's degree is required with particular study in robotics or mechanical engineering and some knowledge of manufacturing processes. Due to the varied working conditions, a field service technician must be able to stand for prolonged periods of time and carry/lift up to 40 lbs. A field service technician interacts directly with customers and clients. Therefore, solid customer service skills are also a benefit.

Good troubleshooting involves knowing how the system operates and how to effectively use test equipment. Most importantly, be patient and systematic.

Troubleshooting Charts

Most manufacturers of robotic equipment provide troubleshooting manuals with charts and diagrams to aid in repairs. Use this information to help locate the faulty module or part. A typical symptom list itemizes problems associated with a specific robotic system, **Figure 11-2**. Detailed information related to the symptom is provided to help identify the problem, **Figure 11-3**. A flow chart may also be provided to guide you to the problem, **Figure 11-4**.

Understanding how the circuit, device, or system functions and knowing how to properly use test equipment are important factors in successful troubleshooting and testing. This is true for both a simple circuit and the most complex system. Practice is the best way to learn the procedures.

11.2 General Servicing Techniques

The individual parts on robots seldom require servicing. Repairs to complex robotic systems are commonly performed on the subassemblies. Individual parts may not be repaired; an entire component or module may simply be replaced.

Figure 11-2. This page from a troubleshooting manual lists symptoms and corresponding pages where detailed service information can be found.

Symptom List	Page
Applications 1 through 5 on at the same time	10.10
Axis does not move	10.11
Axis moves erratically or runs away	10.13
Control panel failures	10.15
Data Error (DE) LED on or blinking	10.17
DI/DO failure (Standard)	10.19
DI/DO (Extended)	10.21
Does not home	10.23
Manipulator power does not work (No other indications)	10.25
MTCB indicator failure	10.26
MTCB switches do not work	10.27
Overtime (OT) LED on	10.28
Overrun (OR) LED on	10.29
Overrun detection failure	10.32
Overrun reset failure	10.33
Power Failure (PF) LED on	10.34
Remote operator panel interface failure	10.35
Repeatability varies	10.36
Servo Error (SE) LED on	10.39
Servo Error (SE) LED blinks and operator panel beeps	10.40
Transmission Error (TE) LED on or communication error	10.42

Figure 11-3. Troubleshooting manuals are an excellent resource for service technicians. A—Detailed troubleshooting information helps to identify the actual problem. B—Manufacturers usually provide complete directions for diagnosing and solving problems. Note the safety warning included as part of a diagnostic procedure.

Defective Battery

NOTE: Do not dispose of the battery in a fire.
With the system off, check the voltage across the battery. The voltage should be between +3.2V and +4.2V dc.
  A discharged battery may load down the power supply while charging. To test for this, unplug the battery and reload the applications.
  If the battery is suspected of only being discharged, leave the controller powered up for eight hours to sufficiently charge the battery. If the voltage is still low after charging, replace the battery.

Low Power Supply Voltages
Check for 24V dc, –12V dc, 12V dc, 5V dc at the power supply. If any of the voltages differ by more than 5%, adjust or replace the power supply. See power supply adjustment (4009). **NOTE:** The power supply may be left powered on with no load attached.
Defective Motor Control Board
If the battery is good and the voltages are good, suspect the Motor Control Board.

A

AXIS DOES NOT MOVE
Defective Servo Pack Fuse
Swap the servo pack fuse for the failing axis. Do not rely on the indicator. Theta 1 is labeled SV1 in the controller. (15 amp) Theta 2 is labeled SV2 in the controller. (10 amp) Z-Axis is labeled SV3 in the controller. (7.5 amp) Roll is labeled SV4 in the controller. (7.5 amp)
Defective Z-Axis Brake
The brake should remain locked until manipulator power is brought up. Then the Z-Axis servo pack takes over by controlling the power to the servo Z-Axis.   If the brake remains locked after manipulator power is brought up, suspect the magnetic brake or circuit to the brake (brake is approximately 85 ohms when good). Also, the brake can be tested by disconnecting connector CN16H with manipulator power on, which makes a clicking noise.
Out of Adjustment Zero
Perform the "Servo Pack Zero Adjustment."
Defective Motor Control Board

1. If only the Z-Axis does not move, set the controller power to 0 (off), then ground CN16H-1 (WD16) to release the brake. Set the power switch to I (on). Move the axis to check for binds. If the axis cannot be moved, isolate the bind. Remove the jumper.
    For any other axis, set the power switch to O (off). Move the axis to check for binds. If the axis cannot be moved, isolate the bind.
2. Set the controller power switch to I (on). Press the Manip Power key on the control panel.
    DANGER: The next step requires you to push on the axis with manipulator power on. Do not allow your body to enter the work space.
3. Attempt to push or turn the axis. If the axis moves easily, then suspect the Servo Pack or wiring from the Servo Pack to the motor (WD07-WD10).
  >>>Continued on the next page.

B

Figure 11-4. A flow chart is a step-by-step guide for diagnosing and correcting problems.

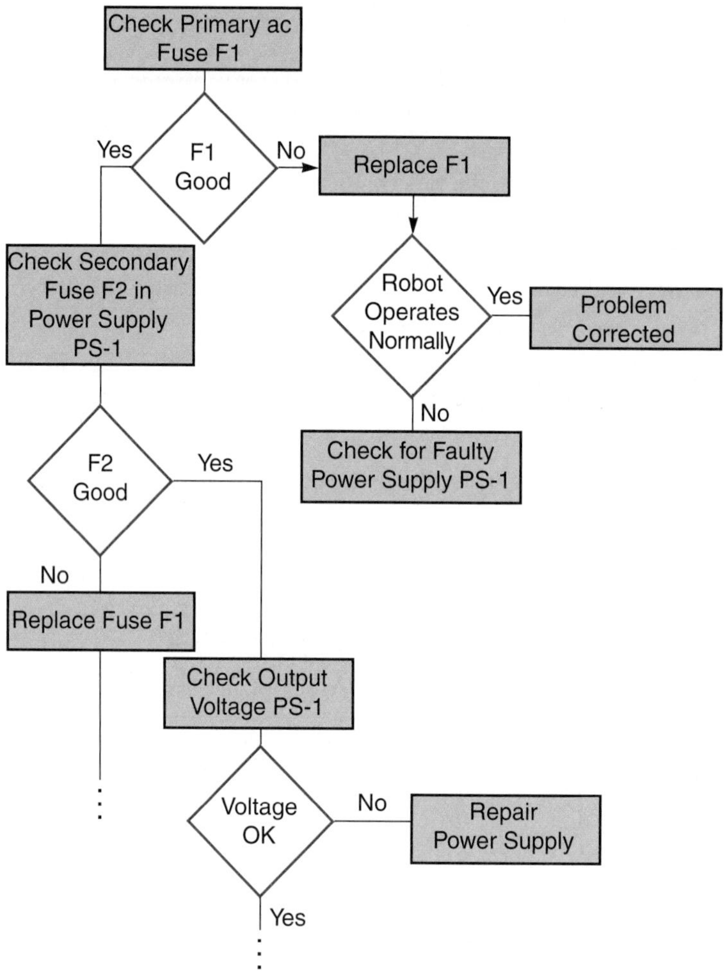

To determine the source of problem, the technician should gather information about the equipment and its operation, evaluate the data to determine a course of action, make the repair, and keep comprehensive records for reference. The seven steps that follow address general servicing techniques.

Step 1. Initial Observation and Inspection

Never accept another person's word about the robot's performance—check it yourself! Determine if the problem is electrical or mechanical in nature. Inspect the power system. Determine if the proper current (single-phase ac, three-phase ac, or dc) is being used to power the equipment. All magnetic contactors, transformers, switches, circuit breakers, fuses, and other electronic controls should be checked.

Observe the physical condition of the equipment. Look for fluid or air leaks, unusual noises, odors, poor electrical connections, or loose wires.

Step 2. Gather Information

If possible, acquire a schematic, a service or maintenance manual, and an operating manual for the robot. These items are essential in servicing most equipment. If these resources are not available, consult with someone who has experience servicing the robot. An identical robotic system may be used to determine specific values (such as voltage, current, and waveforms). These values can then be compared to those in the robot being serviced.

Step 3. Analyze the Equipment

After gathering the necessary information, you should have an understanding of how the system functions. Divide the various components into categories according to purpose. Categories may include input power systems, rectifiers, transformers, filters, feedback loops, amplifiers, and interface/control circuits. Then, identify the input and output of each category.

Step 4. Isolate the Problem

Trace the malfunction to a specific subsystem within the equipment. Use a voltmeter, an oscilloscope, or a multimeter to read measurements or waveforms (**Figure 11-5**). To be successful in finding the problem, you must know how to use the test equipment and how to interpret the measured values. Compare your readings to those indicated on the schematic or in the service manual. Comparing the values of test readings to the recommended values usually isolates the problem.

Verifying a malfunction is sometimes difficult. If the input to a subsystem is correct and the output is not, it may seem logical that the subsystem itself is the problem. This is not true in all cases. The output of one subsystem can become distorted because of another subsystem to

Figure 11-5. The readings from test equipment assists in isolating equipment problems. A—Voltmeters are portable and have various attachments to accommodate different types of equipment. B—This oscilloscope provides a waveform display of the signal sampled. (Tektronix, Inc.)

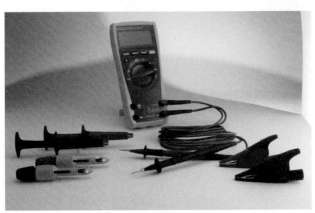

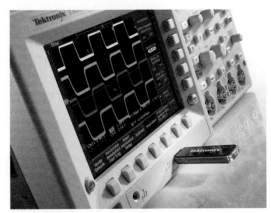

A   B

which it is connected. When this is suspected, disconnect the subsystem you are inspecting and check the voltage and waveforms again.

Step 5. Narrow the Problem to a Component, Circuit, or Module

Take a series of measurements and compare the results to isolate the malfunction to a specific component, circuit, or module. Check the component, circuit, or module to identify the problem. Use the instruments and methods appropriate for working with the malfunctioning unit.

Step 6. Replace the Component or Device

Any replacement component must be equal to or better than the original. Important factors to consider are voltage, maximum current, and power rating. Always replace a component with one of identical characteristics.

Desoldering is also an important step in replacing faulty components. Select the proper iron and soldering tip for the repair and be mindful of nearby components while soldering. Using a *solder sucker* safely removes unwanted solder from a printed circuit board without damaging the board or other components, **Figure 11-6**. Carefully solder the new component into the circuit to avoid damage by overheating. You must also guard against "cold" solder joints, which are not connected accurately and do not allow proper electrical current to flow. From an electrical standpoint, a "cold" solder joint causes high resistance in a circuit. Always use *resin-core solder* (*not* acid core) for *all* electronic circuit work. Resin-core solder creates joints that are noncorrosive and do not conduct electricity.

Step 7. Check the Repairs and Keep Records

After the repair is complete, operate the system for several minutes to be sure it continues working properly. When you are satisfied that the system is functioning properly, document the original problem, what caused the

Figure 11-6. Both powered and spring-loaded solder suckers are effective desoldering tools.

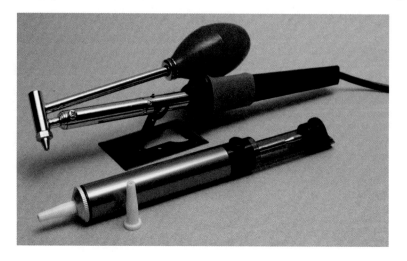

malfunction, and how it was corrected. Keep this information with the service manual or in a separate file, so it is on hand for future reference.

11.3 Preventive Maintenance

Some companies practice management by crisis—they avoid thinking about problems until one occurs. In terms of maintenance, this means equipment is fixed only when it breaks down. Companies that operate this way leave the production line constantly vulnerable to equipment failure.

Equipment failures are costly and affect all areas in the production process. Production bottlenecks cause work in process to back up at work stations, which means the operation must run additional hours to "catch up." This also causes additional production time for the operations that follow. When machines are not in use, idle time results for the operators. Inventory costs increase when materials must be held beyond the normal time it takes to cycle them through the manufacturing process. If an order is not produced in time, the customer may go to another source. If the equipment experienced excessive wear before breaking down, increased rework and scrap can result. *Rework* is the process of fixing parts that no longer meet product specifications. *Scrap* is an equipment part that cannot be fixed. All of these factors result in lost revenue for the company.

Preventive maintenance is a scheduled plan of maintenance tasks that are performed to prevent equipment breakdown and loss of production. This is the best solution to avoid the costly results of equipment failure. Preventive maintenance should be designed to maintain and improve the reliability of equipment by replacing worn components before they fail. Some of these activities include inspecting equipment, changing fluids, lubricating parts, and machine/component refurbishing or replacement at specified periods.

Documentation is key to preventive maintenance. Maintenance personnel can observe and record equipment deterioration to identify the best parts replacement or repair schedule. Preventive maintenance can play an important part in an organization's management strategy by providing a number of long-term benefits, such as improved system reliability, decreased cost of replacement, decreased system downtime, and improved inventory management for spare parts. When the process is closely examined in terms of cost and system performance, preventive maintenance is preferred over performing maintenance only when the system fails.

Preventive maintenance also benefits the daily performance of robots and other types of industrial equipment. It can increase flexibility, help maintain production flow, and allow for a continuous analysis of equipment. Additionally, preventive maintenance decreases the need for a large on-call maintenance crew. When used in combination with mathematical analysis, it aids in predicting when equipment needs to be adjusted or replaced to avoid breakdowns. Preventive maintenance can be performed at convenient times, such as when a robot is not actively involved in a process. This reduces overtime costs for skilled operators and balances the workload among shifts. When performed conscientiously, preventive maintenance ensures consistent quality, reduces product costs, and reduces downtime of critical production equipment.

Developing a Maintenance Program

Developing a preventive maintenance program can be achieved by following a few basic steps.

Step 1. Establish a Schedule

Each shift must share in maintenance responsibilities. A maintenance schedule can be altered to accommodate the appropriate number of shifts, **Figure 11-7**. However, it must be followed carefully for maximum effectiveness. If one shift fails to follow the schedule, it can cause problems for the entire program. The schedule should be easy to maintain and should be posted directly on the machines.

Step 2. Use an Assignment Sheet

An assignment sheet describes tasks that must be performed on a regular basis, **Figure 11-8**. Each assignment sheet identifies the unit to be maintained, the maintenance intervals, and the type of action required. The steps may be carried out in any order, but each should be completed thoroughly. Assignment sheets should also be posted directly on the machines.

Step 3. Keep Work Areas Clean

Once a preventive maintenance program is initiated, machinery and work areas must be kept clean. This allows the operator to readily identify any leaks, visible damage, or wear.

Figure 11-7. A preventive maintenance schedule helps identify and eliminate small problems before they become large ones.

Daily/Weekly Preventative Maintenance Schedule

Machine Type: _IBM-7547_
Department: _179_
Serial Number: _257698ZMW_
Building Number: _5_

Shift	Week	Mon.	Tues.	Wed.	Thurs.	Fri.	Weekly	Operator	Comments
Shift 1	1	X	X	X	X	X	X	John Wasson	Air leak
Shift 2	1	X	X	X	X	X	X	Stephen Fardo	Bad relay
Shift 3	1	X	X	X	X	X	X	Robert Towers	No problems
Shift 1	2	X	X	X	X	X	X	Tim Ross	No problems
Shift 2	2	X	X	X	X	X	X	Stephen Fardo	Air leak
Shift 3	2	X	X	X	X	X	X	John Wasson	Bolt loose
Shift 1	3	X	X	X	X	X	X	Robert Towers	Motor noise
Shift 2	3	X	X	X	X	X	X	Tim Ross	No problems
Shift 3	3	X	X	X	X	X	X	John Wasson	Bad I/O port
Shift 1	4	X	X	X	X	X	X	Stephen Fardo	Loose clamp
Shift 2	4	X	X	X	X	X	X	Robert Towers	Air leak
Shift 3	4	X	X	X	X	X	X	Tim Ross	No problems

Figure 11-8. Assignment sheets should be posted on the machines to identify maintenance tasks and the interval at which they should be performed.

IBM-7547 Preventative Maintenance Assignment Sheet			
Unit	Action	Interval	Lubrication
Manipulator	Clean Check Oil Levels Check for Air Leaks	Daily Daily Daily	— — —
Controller	Check Air Filters	Monthly	—
Roll Axis Belts	Check Tension	Monthly	—
Roll Axis Gears	Lubricate	Monthly	Molykote G
Theta 1 Axis	Change Oil	Semi-Annually	Number 10 Oil
Theta 2 Axis	Change Oil	Semi-Annually	Number 10 Oil
Bearings	Pack Bearings	Every Replacement	Number 23 Grease

Step 4. Make the Operator Part of the Program

Preventive maintenance places a great deal of responsibility on the operator. The operator must be dedicated to the program for optimum success. This does not mean that operators are personally responsible for maintenance, but they must keep maintenance personnel informed of any problems.

While a good preventive maintenance program definitely reduces the frequency of unexpected equipment failure, such programs do have drawbacks:

- Greater need for dedicated workers.
- Increased responsibility and paperwork for operators.
- Allotted access time required to work on the equipment.
- Initial cost of implementing a preventive maintenance program.

Implementing a New Program

Implementing a new preventive maintenance program involves a number of steps, **Figure 11-9**. Operators should be involved early in the planning process because they are so important to the success of preventive maintenance schedules and activities. A good preventive maintenance program can keep essential equipment functioning properly.

Figure 11-9. Five necessary steps in implementing a new preventive maintenance program.

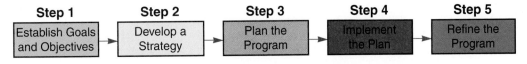

Review Questions

Write your answers on a separate sheet of paper. Do not write in this book.

1. Identify several areas of technology used by robotic systems.
2. What does the term "troubleshooting" mean? Which of your senses is used in troubleshooting equipment?
3. What are some skills that technicians find helpful in troubleshooting?
4. Why are equipment troubleshooting manuals important?
5. List the seven steps involved in servicing a robotic systems.
6. During the initial observation of equipment to be serviced, what should you look for?
7. How does a technician proceed in gathering information about equipment?
8. What are some types of measurements that can be taken to check equipment?
9. How do you narrow down the cause of an equipment problem?
10. What factors should be considered when replacing a component?
11. Why is keeping good records important in servicing equipment?
12. Why is preventive maintenance important?
13. What do the terms "scrap" and "rework" mean?
14. List the steps to follow in developing and implementing a preventive maintenance program.
15. What is a preventive maintenance assignment sheet?
16. What are some disadvantages of preventive maintenance?

Learning Extensions

1. Perform an Internet search to find information on preventive maintenance programs and other robotic maintenance tasks. Prepare a brief presentation on one of the programs, schedules, or methods you found to be the most interesting.

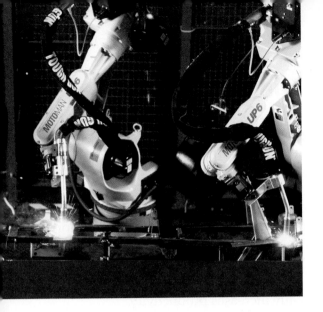

Chapter 12
Robots in Modern Manufacturing

Outline

12.1 Using Robots in Manufacturing
12.2 Evaluating Potential Uses for Robots
12.3 Preparing an Implementation Plan

Objectives

Upon completion of this chapter, you will be able to:

- Identify working environments that are candidates for robot implementation.
- List non-economic justifications for investing in robots.
- Prepare a robot implementation plan.

Technical Terms

avoidance costs
capital investment
cost savings
engineering economics
payback period

Overview

Today, manufacturing is customer-driven. The manufacturing process begins by defining the customer's needs. A company must determine how to meet those needs to keep the customer satisfied, while maintaining satisfactory profit levels. A successful manufacturing process demands a continual effort toward quality, cost-effectiveness, shorter lead times, good customer service, and timely response to changing market conditions. This chapter addresses some business decisions related to the use of robots, the advantages and disadvantages of implementing robotics systems in a manufacturing environment, and how to develop an implementation plan.

12.1 Using Robots in Manufacturing

During the mid-eighties, manufacturers believed that robotic technology was a key element in the factory of the future. Today, many major companies that initially invested millions in robotic systems tell stories of failure, as well as success. Robots and other complex automation systems did not solve every manufacturing problem. In fact, when faced with the prospect of re-tooling an entire line, many organizations scaled back the level of automation and took a more simplified approach.

For some organizations, the following incorrect assumptions produced costly mistakes:

- Incorporating robots automatically increases productivity.
- Intelligent machines can do the same job as trained workers and save labor costs by replacing workers.
- A large investment in robotic automation pays off in time.

Productivity

Does robotic technology automatically increase productivity? No. To date, no machine has been invented that is more flexible or more efficient than a motivated human worker equipped with the right tools. In certain applications, robots can be more productive than humans over a long period of time because robots do not suffer from fatigue and do not require downtime for food or rest. However, productivity does increase when the product being manufactured is specifically designed for automated assembly. However, manual assembly processes benefit from specific design considerations as well.

Replacing Employees

Many companies assumed that robots would replace a significant number of employees and the resulting savings would offset the initial cost for the robotic equipment. When robotic systems were implemented to replace production workers, computer programmers and other highly trained technicians were hired to program and maintain the robotic systems, **Figure 12-1**. As a result, companies were left with fewer workers who understand the manufacturing processes.

In general, direct labor costs are not a major portion of the total cost in producing a product. Typically, a robot replaces less than 2 workers in an industrial process or application, which results in little direct labor savings, **Figure 12-2**. In fact, adding robots and other high-tech equipment can increase direct labor cost due to the additional skilled individuals required to support the technology.

Chapter 12 Robots in Modern Manufacturing 267

Figure 12-1. Maintenance technicians and programmers are needed to ensure the robot correctly performs the tasks required and operates efficiently.

Figure 12-2. Direct labor costs comprise about 15 percent of the total cost in producing a product.

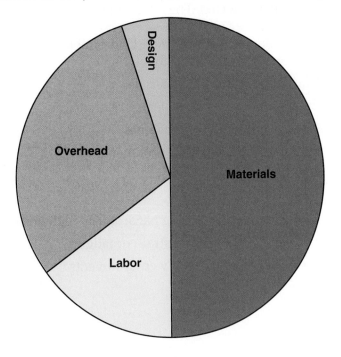

The Pay Off

Do large investments in robotic systems pay off over time? Not necessarily. After spending millions of dollars on such systems, a company may be forced into long production runs to pay for the equipment. A simple downturn in the economy can destroy such a plan overnight. Additionally, manufacturing methods have changed. Industry trends, such as fast product turnaround, can add significant costs to a production process that relies on robotic systems. Re-tooling automated manufacturing systems is expensive. If robots or other automated systems cannot be changed or updated economically, a large investment in these systems is not worthwhile.

Successful Robotic Implementations

The automotive industry has made a large investment in robotics due to the very nature of the business, **Figure 12-3**. Global marketing, worldwide competition, and attempting to meet a wide range of customer preferences have resulted in numerous products and models offered each year. Shorter product development time and less changeover lead time motivated the automotive industry to turn to robotic automation. To accommodate the robotic systems, task complexity was reduced by altering product and fixture designs. Production line flexibility has been enhanced by the use of programmable robots.

Undesirable Environments

Robotic installations are most easily justified in undesirable environments. Whenever working conditions are unpleasant or unsafe, worker productivity is affected. The use of robots in such environments has proven to be an effective solution. If working conditions cannot be improved, the operation should be automated. Undesirable conditions suitable for robot implementation include the following:

- Handling hazardous materials.
- Environments that lack clean, breathable air.
- Operating hazardous machines.
- Extreme temperature environments.
- Handling heavy materials.
- Areas with poor lighting.
- Environments containing hazardously high noise levels.
- Tasks that cause mental strain or fatigue.
- Excessively dusty or dirty environments.

Figure 12-3. A comparison of the use of robots in the manufacturing and automotive industries. Notice the predominance of robots in the automotive industry. (Adapted from data in the *World Robotics 2004* survey by the UNECE.)

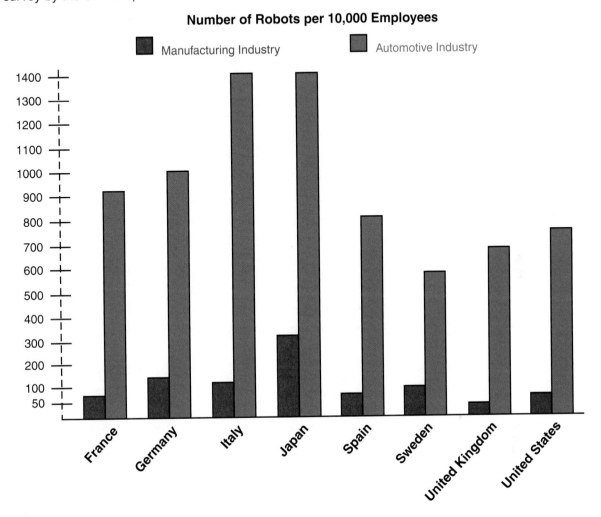

12.2 Evaluating Potential Uses for Robots

Most robotic applications cannot be simply defined. A potential application can be evaluated by answering five questions:

- Does the technology match the need?
- Have all advantages and disadvantages been considered?
- What are the non-economic justifications?
- What are the economic justifications?
- Is the organization ready for robots?

Matching the Technology to the Need

When evaluating a robotic system, consider the production characteristics of the process. An automation system may be beneficial if production volume ranges from medium to high, production cycles are short- to medium-length, lot sizes are small, and several workpieces are produced per process. The need for material and product handling should also be reviewed. Robotic systems can be an effective addition when the process requires handling heavy loads, precise positioning of materials and products, and general (not specific) orientation of parts.

A process is not a candidate for implementing robotic systems when one or more of the following conditions exist:

- Critical aspects of the workpiece or the process cannot be measured.
- Specific workpiece and process standards do not exist.
- Identification of the workpiece is difficult
- Positioning the workpiece is complicated.
- Evaluating the condition of the workpiece is intricate.
- The range of movement needed for the robot in the process cannot be readily identified.
- The robot's movement cannot be contained within a limited space.

Considering the Advantages and Disadvantages

When determining whether a robotic implementation is justified, carefully consider and evaluate the advantages and disadvantages involved in making the change. The balance of advantages and disadvantages shifts with each unique environment and situation.

Advantages

Robots can offer increased productivity, given the right conditions. While robots are no faster than humans, the increased productivity results from the robot's constant work pace. During an eight-hour shift, the robot usually outperforms a human operator—especially if the task is repetitive, boring, heavy, or performed under poor working conditions. Further benefits to productivity can be accomplished by allowing the robot to operate for more than one shift.

Robots can also improve product quality because of their accuracy and repeatability. For example, a robot used in a spot welding application may be able to place a weld within 0.050" (0.127 cm), time after time. This accuracy is hard for a human to duplicate, since the spot-welding gun weighs approximately 100 lbs. (45 kg). Even though the gun is counterbalanced, fatigue takes its toll after several hours, **Figure 12-4**.

Figure 12-4. A robot's performance is not affected by fatigue. A—During a typical work shift, a human operator tires from handling and operating heavy welding equipment. B—A robot does not experience any strain while spot welding an automotive body assembly during a single work shift, or longer.

A

B

Automated equipment requires consistently uniform components to function efficiently. This requirement adds to the improvement of product quality, **Figure 12-5**. A human operator may be able to adjust the process and assemble components that are a bit out of spec. Automated equipment, however, cannot compensate for irregularities in the product. Parts must be within programmed specifications and are held to closer tolerances. To take full advantage of automated production, manufacturers demand 100 percent quality parts from vendors.

Installing robots reduces personal injuries and increases employee safety. The passage of the Occupational Safety and Health Act (OSHA) provided an incentive for manufacturers to introduce robots in many operations that were considered dangerous. Using robots in these areas reduces the need for complex safety procedures and equipment. Automating unpleasant or dangerous tasks also increases worker morale. If workers are moved from dirty, hostile, and hazardous environments, their attitude toward working improves and they can be placed in more challenging positions with greater responsibility.

Robots can contribute to holding down or reducing production costs. Some manufacturers boast of productivity gains of up to 400 percent. Others have reduced scrap material and time spent reworking products. For example, General Motors once reported that four-fifths of rejected products in spray painting came from manual operations. Reducing the amount of

Figure 12-5. To function efficiently, assembly robots must work with components of consistently high quality. The result is a final product of higher quality. This robot is assembling electronic printed circuit boards.

material used also helps reduce production costs. Using robots for spray painting can result in a 50 percent reduction in the amount of paint used. Additional savings may result from the reduction of energy required to light, heat, cool, or ventilate work areas where robots operate.

Most automated equipment is dedicated, or designed to perform one function. Trying to adapt dedicated equipment to new operations often proves costly or difficult to accomplish. Robots offer the advantage of flexibility; they can be reprogrammed to perform new tasks. An example is a robot's ability to perform operations on different automobile models on the same assembly line. Additionally, manufacturing operations are typically performed using groups of 50 or fewer parts. Robots are well-suited for operations of this type.

Disadvantages

The cost of implementing robots in manufacturing environment is a primary disadvantage. The money that a business spends to purchase fixed assets, such as buildings and equipment, is known as a *capital investment*. The addition of robots, like other pieces of equipment, must be economically justified. However, the decision should consider factors such as worker health and safety.

Using robots directly impacts the production line. For example, parts must be properly positioned and oriented for use by the robot. The manufacturing workflow changes and additional work space is required. After a robot is acquired, all the tooling and equipment required to interface the robot with the manufacturing line is necessary. This tooling and peripheral equipment is very expensive and, depending on the application, the cost of the equipment can exceed the cost of the robot itself.

Another disadvantage is employee opposition. Some workers feel threatened by robots and need to be reassured. If management and workers do not develop a positive attitude regarding the use of robots, installation and implementation will most likely fail.

Non-Economic Justifications

Non-economic considerations sometimes outweigh economic ones. The use of robots in hostile environments is an example. If poor working conditions are ignored by management, productivity can decrease and product quality may suffer. Workers may become genuinely ill or may call in sick to avoid unpleasant conditions. The cost of lost production can far exceed the cost of using robots.

Looking at the long-term impact of using robots may justify the expense today. Over a period of years, will the use of robotic technology:

- Produce better products of more consistent quality?
- Reduce new product lead time?
- Increase flexibility of the production line?
- Improve customer service and product satisfaction?
- Improve market response time?

If the answer to several of these questions is "yes," an analysis should be done to determine if the capital investment is justified. However, other alternatives should also be explored. One alternative may be to modify or rebuild present equipment. The same goals may also be accomplished through retraining and motivating the workforce.

In a highly competitive market, new technology may help maintain or improve a company's competitive position. See **Figure 12-6**. This, too, may justify investing in robotic equipment. Robots are often used in research and development. The expertise gained in this area can be applied later to their general use. Economic returns occur because the time needed to get robots up and working is reduced.

Sometimes it takes competitive pressure to initiate change. If one competitor starts to use robots and gains a price advantage, robots may become more attractive to other companies in that market. However, buying a robot should be based on firm business factors, not simply to follow trends.

Economic Justifications

Even though the previous factors are considered non-economic, each makes an economic contribution. But, the decision to invest in robots may also be based solely on economic factors. *Engineering economics* involves analyzing the economic impact and technical aspects of engineering projects. All the factors involved in purchasing, implementing, and maintaining robotic equipment for a manufacturing business are analyzed in this process.

There are two situations where economic analysis can assist in making an investment decision. The first is investing in equipment for a new application, which will prevent additional costs in the future. This type of investment is referred to as *avoidance costs*. The second is replacing an

Figure 12-6. The robotic manufacturing and assembly equipment used by the Nissan Motor Company allows them to change, update, and expand vehicle model and body options in order to meet consumer demand and stay competitive in the automotive industry.

Impact of Robotic Equipment on Auto Manufacturing and Assembly Operations		
Production Year	Models	Body Types
1970	8	10
1990	27	54
2005	11	17

existing production or assembly method to reduce current costs. This type of investment is known as *cost savings*.

Investing in a robot must provide a return equal to or greater than the required capital. Most companies have established benchmark criteria for return on investment and *payback period* (time required to recover the amount spent for new equipment, either through savings in labor or materials). For example, a company may require a 25 percent return on investment and a payback period of less than two years. In addition to return on investment and payback period, factors considered include discounted cash flows, depreciation, tax rates, present value, net present value, cost of capital, and production.

Is the Organization Ready for Robots?

A critical question for many companies is, "Is the organization ready for robots?" This is not a simple question, since many variables are involved in making a judicious decision. Robots can do the same work as humans, but accomplish the work more efficiently. Robots never get sick or need rest, and they can work "24/7." Robots perform precise operations on repetitive work at a rapid rate.

On the other hand, robots do not have the capacity to be innovative or collaborative and cannot think independently. Robots do not learn from their mistakes or have the ability to make major changes and independently adapt without help from a human operator. Once the technology, workflow impact, and financial aspects have been researched and discussed, a decision can be made about the implementation of robots. But, this decision can only be made if the organization is ready for robots!

12.3 Preparing an Implementation Plan

After deciding that using robot equipment in a manufacturing process will help the company reach its production and business goals, the following steps can be used to develop a successful implementation plan.

1. Identify potential applications.
2. Analyze potential applications.
3. Construct a matrix for comparing equipment.
4. Review the available equipment.
5. Match the robot to the application.
6. Develop the proposal.
7. Develop and refine the application.
8. Begin implementation.
9. Provide training.
10. Provide maintenance programs.

Identify Potential Applications

One way to identify possible applications for robotic equipment is through a plant survey. Certain job, product, or process characteristics can be used to identify potential uses for robots. Some general areas to begin the survey include material handling, component insertion, inspection, and testing. Potential applications can also be identified by reviewing personnel and safety records. For example, if a job is monotonous, dirty, hot, noisy, or presents a possible health or safety hazard, it is a candidate for a robot. Other characteristics of manufacturing jobs that may be performed by robots include short product life, frequent design changes, using families of parts, using a minimum of tools and parts, and operating for multiple shifts.

Analyze Potential Applications

Once possible robotic applications are identified, select the most promising applications and construct profiles of each. This typically involves several trips to the production floor and talking to people connected with the operations. The importance of drawing on the expertise of operators, supervisors, engineers, and other manufacturing personnel cannot be overemphasized.

Installing a robot can have a ripple effect on the whole production line. For example, additional material handling and feeding devices may be required. Blueprints, specification sheets, production control records, quality records, and safety records must all be explored to create a complete picture of the production process. Compile data on the number and sequence of operations, parts flow, cycle times, product volume, personnel requirements, product life, results of quality checks, and the effects of environmental factors. This information is useful in determining if a robot is really capable of handling a job.

Construct a Matrix for Comparing Equipment

Matching a robot to a particular manufacturing application is not an easy task. To evaluate various robots, construct a matrix to simplify the comparison of key elements (**Figure 12-7**). The matrix should include categories for the model number, price, number of axes, type of controller, load capacity, work envelope, type of power supply, speed, accuracy, repeatability, and methods of programming.

Review the Available Equipment

Research all of the available products that meet the requirements of the selected applications. Review information about the capabilities and limitations of various models. Information can be acquired from robot manufacturers, as well as the Internet. In researching various products, you can become familiar with common elements, as well as model-specific features and packages. Compile the data from this research into the matrix for review.

Figure 12-7. This sample matrix presents the features and capabilities of three robotic systems.

Equipment Comparison Matrix			
	System #1	System #2	System #3
Manufacturer	Motoman	FANUC	ABB Robotics
Model Number	UP200	M 420iA	IRB 6600
Horizontal Reach	2951 mm	1855 mm	2550 mm
Number of Axes	5	4	6
Controller Type	Teach Pendant	Teach Pendant	Teach Pendant
Load Capacity	200 kg	40 kg	225 kg
Mass	1350 kg	620 kg	820 kg
Structure	Vertical Jointed Arm	Articulated	Vertical Jointed Arm
Repeatability	+/- 0.2 mm	+/- 0.5 mm	+/- 0.1 mm
Programming Method	Teach Pendant	Teach Pendant	Teach Pendant
Other Features	Material Handling	Material Handling	Material Handling

Match the Robot to the Application

At this point, the task is to determine if the job can actually be handled by a robot and develop a method for using the robot. In addition to the robot itself, tooling and peripheral equipment must be assessed. The robot's cycle time should also be established to further match the robot with the application. If laboratory facilities are available, construct a prototype to establish cycle times and evaluate robot performance and the process sequence.

Develop the Proposal

A proposal is an all-encompassing document. It begins by stating the problem and proposes a solution supported by rationale. The rationale should include both economic and non-economic factors. The proposal addresses personnel requirements, required resources, schedule, and budget. Additionally, a financial analysis is prepared that reviews investment, expense items, savings, and other economic factors to determine the most efficient distribution of capital. Return on investment, payback periods, and discounted cash flows should be considered for each alternative. Proposals should always contain economic justification.

Develop and Refine the Application

After the proposal is approved, the details of the plan are itemized. The necessary tooling and safety devices are designed and ordered. The specifics of the robot application are refined and debugged. Much of the time spent on the project is devoted to this area. This is the initial proving ground for the application.

Begin Implementation

During implementation, the site is prepared for production. The floor space is prepared and the various service drops and safety devices are installed, **Figure 12-8**. While a service drop usually consists of an electrical line run to a piece of equipment, it can also include air, water, or any other utility necessary to the equipment operation. Simply plugging a robot into a production line is not possible. All the analysis, planning, and preparation up to this point should ensure a successful implementation process.

Provide Training

During implementation, everyone involved in the operation needs training prior to actual installation. Maintenance workers may already

Figure 12-8. Safety fences must be installed around the robotic work station during the implementation phase of the project. (Cisco-Eagle)

possess some of the technical skills needed to work on robots. However, they need to become familiar with the manufacturing process and the basics of programming. Robot manufacturers usually provide training for maintenance and other personnel, either in their own facilities or at the installation site. In-service and retraining programs are also a vital part of any manufacturing operation. These programs are needed not only for technical personnel, but for all employees. Identify the various groups involved in the manufacturing process, so training can be tailored to meet the needs of each group.

Provide Maintenance Programs

Establishing a maintenance program is important to success. Hire an adequate number of maintenance personnel. Consider that workers become sick, go on vacation, are transferred, retire, or accept positions with other companies. A maintenance program also includes stocking an ample quantity of spare parts. This allows repairs to be made quickly, which avoids loss of production time. Robot manufacturers supply a list of spare parts to stock. As experience is gained with the robot, compile a list of parts that applies to its specific operation.

Review Questions

Write your answers on a separate sheet of paper. Do not write in this book.

1. Identify three common assumptions companies make about implementing robots in manufacturing and discuss why each is incorrect.
2. List six examples of undesirable working environments where robots are an effective solution.
3. When evaluating an application for potential robot installation, what are the five questions that should be answered?
4. Robots are generally not faster than humans in performing many tasks. Yet in terms of productivity, a robot (when used to its full potential) is more productive than human workers. Why?
5. How does the use of automated equipment affect the quality of a product? Explain your answer.
6. The decision to invest in robots can be based on non-economic, as well as on economic, factors. Identify and discuss four non-economic factors that can influence the investment decision.
7. What are the ten steps involved in preparing an implementation plan?
8. List five items that are addressed in a proposal.
9. Once robotic equipment has been installed and employees have been trained, what must the company do to ensure successful, long-term use of the equipment?

Learning Extensions

1. Identify a process that would consider the use of a robot.
    A. Identify the advantages and disadvantages of implementing robotic technology into the process.
    B. Provide economic justification for purchasing the robot.
    C. Prepare an implementation plan that addresses each of the ten steps.

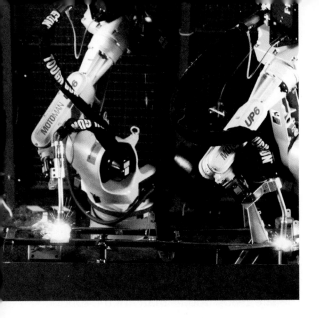

Chapter 13
The Future of Robotics

Outline

13.1 Fully-Automated Factories
13.2 Robots Outside the Factory
13.3 Artificial Intelligence (AI) and Expert Systems
13.4 Impacts on Society
13.5 Your Future in Robotics

Objectives

Upon completion of this chapter, you will be able to:
- Identify current applications of service robots.
- Explain the impact of artificial intelligence on the field of robotics.
- Summarize the difference between expert systems and regular computer systems.
- Describe the impact computer-controlled machinery has on the workforce.

Technical Terms

expert systems　　　　serbots
microbots　　　　　　telerobotics

Overview

The future of any technology is difficult to predict. However, changes resulting from advances in computer applications continue to profoundly affect robotics. Computer-controlled manufacturing has often been referred to as the "second industrial revolution." Uses for robots outside of industry continue to expand, with applications limited only by the imagination. Robots are used in space, medicine, exploration, and countless other areas.

This chapter presents developments that impact the factory of the future, non-factory robots, artificial intelligence, and worker training programs.

13.1 Fully-Automated Factories

For decades, scientists, engineers, and plant managers have dreamed of a fully automated factory without human workers. In such factories, lasers are used to inspect parts and check for wear on machine tools, and assembly components are shaped by automated equipment. Driverless, automated guided vehicles move through the factory picking up and delivering parts and robot arms assemble components the way a human would. This type of operation eliminates complicated fixtures and parts feeders, with only a few humans working among the multitude of robots. The entire automated system can be retooled simply by switching the software program that guides the robots. This fully automated factory model was described in a *Business Week* magazine article in the late 1980s. Today, many of the robotic systems described have been realized and are in use.

The technology for fully-automated, workerless factories exists today. Implementing such technology is not easy, however. Automotive companies have found that simply increasing the number of robots on an assembly line does not necessarily translate into increased sales and profits. The time and cost invested in getting multiple robots programmed and running error-free has been prohibitive for many companies. The hesitation of companies to fully automate production line has been due, in great part, to a misunderstanding of how to use current technology. For example, applying robots to poorly understood production problems can make a bad situation even worse.

For years, American manufacturers lag behind their Japanese and European counterparts in the knowledge and expertise needed to successfully apply robotics technology. George Devol, who patented the first industrial robot in 1954, criticized the "lack of interest in our manufacturing industry and automated systems." He noted that "European and Japanese companies embrace this concept." High degrees of automation are possible with developments and advancements in robot technology, including vision systems, smart sensing systems, and artificial intelligence.

13.2 Robots Outside the Factory

In the 1980s, the robotic industry was in high gear. Hundreds of companies were involved in manufacturing and installing industrial robotic systems. By the early 1990s, however, only a very small number of companies remained actively involved. Major growth outside the industrial sector began in the 1990s with service robotics. *Serbot* (service robot) applications range from mobile robots on building security patrol to automated robotic vehicles that explore other planets. Diverse serbot applications have developed with technological advancements, including:

- Serbots used to clean floors.
- Agricultural applications, such as a robotic "hired hand" that works on a dairy farm. The robot's arms adjust a cow's legs, wash the udders, and attach a milking machine.

- Space exploration serbots designed to examine the surface of Mars.
- Robots are used for education and entertainment by museums and zoos to operate replicas of prehistoric animals.

The field of *telerobotics* has produced robot arms and mobile robots that are operated by remote control (radio transmitter and receiver). These have been used for both deep sea and outer space applications, such as the robot rovers used on the NASA Pathfinder Mission in the late 1990s. The remote controlled rover used in Pathfinder Mission was placed on Mars to transmit data about the surface of Mars back to Earth. Such highly-visible and publicized events have increased interest in remotely controlled telerobotic applications. The use of robotics in the home is a very popular and expanding area of robotic applications. Consumer remote control robotic systems can accomplish several domestic tasks, such as cleaning the floors and mowing the lawn, **Figure 13-1**.

Figure 13-1. Telerobotic products are available to consumers and can make life more enjoyable by handling home maintenance tasks. A—The Dirt Dog® can handle tough floor debris and maneuvers many different surface types. (iRobot Corporation) B—The Looj™ travels through gutters and clears debris and clogs. (iRobot Corporation) C—The Verro™ cleans in-ground pools in a little as 60 minutes and circulates 80 gallons of water per minute. (iRobot Corporation)

Microbots, the tiniest robots, can now perform operations inside the human body. Doctors can guide their work by means of computer control. Imagine a robot submarine small enough to cruise through your veins and arteries! Some people predict that by the year 2029, insect-sized robots will construct, alter, and clean everything from ranch houses to silk suits.

Several manufacturers have developed robots that are used primarily for educational experimentation. These robots are smaller versions of those used in industry, with almost identical operational characteristics. Students in the field of robotics, or related manufacturing fields, can use these educational robots to learn basic operation principles.

Military operations have made use of robots and robotic technology for several years, **Figure 13-2**. Remote-control capabilities provide a safety measure when performing dangerous tasks, such as minefield detection and removal of improvised explosive devices (IEDs). Applications for government and security robots is an area that has received much emphasis in recent years, **Figure 13-3**. Such applications include the handling and disposal of hazardous materials (Hazmat) with robotic systems rather than exposing humans to the risk of exposure and injury.

Figure 13-2. The SUGV (Small Unmanned Ground Vehicle) is a mobile robot that gathers data on its environment and surrounding activities to keep military and security personnel out of harm's way. (iRobot Corporation)

Figure 13-3. Robots used for security tasks must be mobile and versatile. A—The PackBot® can climb stairs, navigate uneven terrain, and is light enough to be transported and positioned by hand. With numerous accessory kits available, the PackBot can be configured to meet a variety of security and inspection needs. (iRobot Corporation) B—In addition to moving payloads of more than 150 pounds and surveillance functions, the Warrior™ robot is also capable of handling bomb disposal operations. (iRobot Corporation)

A

B

13.3 Artificial Intelligence (AI) and Expert Systems

Machines may be developed that are genuinely intelligent and seem humanlike in their problem-solving abilities. "Thinking machines" have long been the goal of scientists and experimenters. However, most researchers agree that we have quite a way to go in developing a truly humanlike thinking machine. Some feel that such a machine can never exist, and ask the age-old question: "Can any mechanism really think?" They believe that a human being is more complex and sophisticated than a machine could ever be. However, this does not mean that machines cannot be made more intelligent than they are currently.

Marvin Minsky is one of the fathers of the artificial intelligence (AI) field and defines it as "the science of making machines do things that would require intelligence if done by man." AI has been called the force behind the "second computer revolution." AI researchers come from many scientific disciplines, including computer science, psychology, cognitive science, philosophy, biology, and engineering. AI research can be broadly classified into three major interrelated areas:

- Robotics. Relates to machine vision, movement, and tactile sensing. The goal is to perform these operations in an intelligent, humanlike manner.
- Natural language processing. Evaluates the "user interface" where human and machine meet. The goal is to develop machines that understand natural human speech.
- Expert systems. Programs that contain a knowledge base, acquired from human experts, which can be used to help nonexperts diagnose problems and make decisions.

For many years, researchers have conducted developmental work on software using AI techniques and related hardware to solve one of industry's most costly problems—lengthy setup and process planning time required to get machine tools ready for a new operation. With the help of AI, a new generation of intelligent machine tools are now able to run with little human help. These machines choose the correct tool, tool speed, cutting fluids, and optimum cutting strategies. The tools can then assemble and inspect the parts, maintaining correct tolerances.

Robotics in Society: Humanoid Robots

Developing robots with the physical characteristics and cognitive capabilities of humans has been a challenge and fascination since the time of Leonardo DiVinci. Humanoid robots are typically created with the human features of two arms, two legs, a torso, and a head. More importantly, a true humanoid robot is also capable of receiving, processing, and responding to information from its environment. From musical performances and verbal communication to walking and maneuvering stairways, advancements in robot technology have led to some amazing humanoid abilities.

Honda's ASIMO robot is a well-known representation of a humanoid robot, as it has made appearances on television talk shows, nationally-televised parades, and been featured at amusement parks. ASIMO can walk, run, climb stairs, and pick up objects. Its cognitive capabilities include reacting to its environment, understanding and responding to simple commands, and recognizing a familiar face.

As humanoid robot technology advances, the robot will likely take on a more human form and develop more complex reasoning and communication abilities. Future applications for humanoid robots may include assisting the elderly or disabled, performing hazardous tasks, or extended exploration missions.

ASIMO (Advanced Step in Innovative Mobility) integrates advancements in robot mobility and intelligence. (American Honda Motor Co., Inc.)

The fastest-growing branch of AI is *expert systems*. An expert system contains a knowledge base, an inference engine, and a computer-human interface. Engineers gather expert information for a specific field and build that knowledge into the knowledge base of the system. The information is retrieved as data by users of the system, or is carried out as logic operations.

Expert systems differ from regular computer programs in a number of ways. Expert systems can reason like a human, can handle uncertainties, and can learn from previous experiences. Additionally, this type of system can explain why it needs further information and how it reached a specific conclusion. Expert systems have come out of the laboratory and have various applications in industry, and in the service and public sectors to solve a range of problems.

One of the first expert systems developed was known as MYCIN. It was developed in the mid-1970s as a medical expert system to diagnose the possible cause and treatment of bacterial infections. Even though it was never put to practical use, the MYCIN system became a model for later reasoning, rule-based expert systems.

Many other applications for expert systems have since been put into practice:

- Automotive expert systems can detect motor problems before an actual breakdown occurs.
- System design applications allow sales people, who are not engineers, to design computer networks for their customers.
- Inventory systems help sales personnel keep track of the thousands of parts required by some complex systems.
- A filler-metal selection system uses AI to recommend optimal filler materials for welding of aluminum.

AI is an important frontier in the advancement of robotic and computer technology, with a growing number of practical applications becoming commercially available. Ultimately, AI could change our way of life.

13.4 Impacts on Society

There is little doubt that robots and other forms of computer-controlled machinery will continue to increase in number and sophistication. To date, few workers have been completely replaced by this technology. However, the employment opportunities for semiskilled and unskilled workers will decrease, **Figure 13-4**. Studies on the impact of industrial robots on job displacement and job creation reached the conclusion that a remarkable skill modification emerges when the jobs eliminated are compared to the jobs created. The jobs eliminated are semiskilled or unskilled, while jobs that are created require significant technical background. Retraining workers is an important activity for every high technology company, if it is to remain efficient and competitive. Currently, computer literacy is an essential employment prerequisite for all workers in the vast majority of job fields.

Figure 13-4. Robots are currently used to perform many manufacturing tasks. A—This robotic system can perform palletizing operations at a rate of up to 20 bags per minute. (Columbia/Okura LLC) B—Dual-grinding operations are performed by this robotic system. (FANUC Robotics)

A

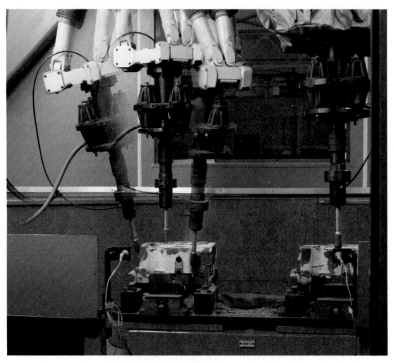

B

Jobs that are highly labor intensive or that pose ergonomic injury potential for human workers are often considered candidates for automation. Robots are used in a wide variety of industries for a broad range of applications, including automotive, foundry, and heavy material handling, packaging, and palletizing. With rising costs, increased competition, and a shortage of skilled workers, companies search for ways to achieve new levels of manufacturing efficiency. This includes using robots for applications that were not previously possible. Vision-guided robotic systems use digital imaging and intelligent software that allow the robots to "see" and "reason." Vision-guided robots can accurately perform inspection, handling, and assembly tasks. This type of robot relies on camera images and can perform multiple operations on the same line without the need to re-tool. This not only increases productivity, but also reduces ergonomic risks.

The future of robotics may be highly dependent on combining computer capabilities and peripheral systems, such as vision-guided technologies and remote control, to develop true human-like machines capable of assisting in homes, offices, and public areas. For example, a humanoid butler could assist disabled people at home and a humanoid porter could carry heavy bags for passengers in airports or train stations. The application of such robotic systems is limited only by the imagination of the designer. Robotic systems with social conscience has energized many people, including students, with the desire to use robots to improve the world.

13.5 Your Future in Robotics

Robotics and automated manufacturing offer careers that can be both exciting and challenging. A variety of educational opportunities are available to students and workers from diverse backgrounds. Specific training is needed for those who operate, maintain, or supervise robots. Training is also needed for workers displaced by automation. Most companies attempt to retrain displaced workers for more satisfying jobs. In some cases, workers may be retrained to program, install, or service robots; operation and servicing is a major industry concern. Although many workers have not been in an educational environment for a long time, well-organized training programs offer a career boost for most workers.

Many of the applications and functions mentioned are provided by suppliers of robotic systems. Training provided by suppliers is usually short and focuses on operation and programming of the systems being installed, with coverage of system maintenance and some special system applications that encourage the sale of products.

Colleges and technical schools also offer programs in robotics. Some focus on operator training, while others develop supervisors for computer-integrated manufacturing. Essential subjects include robotic fundamentals, programming, electricity, basic electronics, hydraulics, pneumatics, digital electronics, microprocessors, programmable logic controllers, and machine vision systems. Basic courses in math, science, and communications skills are also recommended.

Review Questions

Write your answers on a separate sheet of paper. Do not write in this book.

1. Identify several practical developments that have come from the field of service robotics.
2. What is *telerobotics*?
3. Define *artificial intelligence (AI)* and give examples of its use in industry.
4. List the three area classifications of artificial intelligence.
5. What is an *expert system*? Describe how it is used in intelligent machine tools.
6. How is an expert system different from a normal computer program?
7. Explain the skill modification that occurs in jobs that are created when a manufacturing process is automated.
8. What are some of the essential areas of study for a career in computer-integrated manufacturing?

Learning Extensions

1. Visit the Web sites of the robotics laboratories of Stanford University and Carnegie Mellon University. Review the current research projects at each of these universities. How many of the research projects relate to manufacturing and industrial automation?

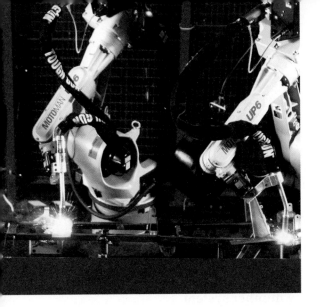

Glossary

A

accumulator
Type of register that stores values for mathematical and logic operations in process. (9)

accuracy
An accuracy measurement expresses how precisely the robot's hand is programmed to reach a predetermined point. The value associated with accuracy is expressed as half of the spatial resolution and is a by-product of command resolution and mechanical inaccuracies. (4)

acoustical proximity sensor
A type of proximity sensor that reacts to sound. Standing sound waves are generated within a cylindrical, open-ended cavity inside the sensor. The presence of a nearby object interferes with these sound waves, which alters the sensor's output. (7)

actuator
A motor or valve that converts power into robot movement. (2)

address register
A register compartment that temporarily stores the address of a memory location that is to be accessed. The address register is programmable in some units, which means that new instructions may alter its contents. (9)

alternating current (ac)
A source of electrical power in which electrons flow in one direction and then change course to flow in the opposite direction. (5)

analog information
Format of computer-processed information that varies continuously. (9)

AND gate
Type of logic circuit that has two or more inputs and one output. (9)

angstrom (Å)
A unit of light measurement. One angstrom equals one-tenth of a nanometer. (7)

anthropomorphic
An object that is in a human form or has human attributes. (1)

arithmetic logic unit (ALU)
A component of a microprocessor that performs mathematical and logic operations. It works automatically using signals sent from the instruction decoder. (9)

armature
The electromagnet in an electric motor that rotates within a magnetic field and creates the motor's torque. (5)

artificial intelligence (AI)
The science and engineering of making machines perform operations commonly associated with intelligent human behavior, such as making decisions based on known information. (1, 3)

automated guided vehicle (AGV)
Computer-controlled, battery-operated transportation devices that operate using one of several navigation options: buried wire guidepath, magnetic tape guidepath (inertial guidance), laser target guidance, or GPS. (4)

automatic tool changer
An automated piece of equipment that can automatically change an end effector, when necessary. (8)

automaton
A human-made object that moved automatically. Originally used for what we now consider to be a robot. (1)

avoidance cost
An investment made in equipment or other business/production assets that is projected to prevent additional costs in the future. (12)

B

bifilar construction
A method of winding stator coils in which two separate wires are wound into the coil slots at the same time. The two wires are small, permitting twice as many turns as with a larger wire. (5)

binary-coded-decimal (BCD) number system
A system of counting in which four binary digits are used to represent each decimal digit; devised to simplify the use of large binary numbers. (9)

binary counter
A device used to count numeric information in binary form. (9)

binary logic circuit
An arrangement of bistable devices that makes logical decisions based on input signals. (9)

binary number system
A number system with 2 as its base. The largest value that can be expressed by a specific place is the number 1, therefore, only the numbers 0 or 1 are used. (9)

binary point
The binary equivalent to a decimal point in a base 10 number system. (9)

bistable
Describes an electronic device that can be set in one of two operational states by an outside signal and can store one binary digit or bit of information. (9)

bit
A single pulse of voltage processed by a computing system. Also called a *binary digit*. (9)

brushes
Carbon devices that rub against the commutator in an electric motor. When power passes through the brushes and commutator on its way to the armature, additional magnetic fields are created. (5)

bus network
A series of registers connected together. (9)

byte
A group of 8 bits. Also called a *binary word*. (9)

C

capacitance
A. The ability of a material to hold an electrical charge. B. The ratio of charge on a conductor to the potential difference between conductors. (7)

capacitive transducer
A type of transducer that measures a change in capacitance. (7)

capital investment
The money that a business spends to purchase fixed assets, such as buildings or equipment. (12)

Cartesian configuration
The movement of a robot along three intersecting, perpendicular, straight lines; referred to as the X, Y, and Z axes. (2)

central processing unit (CPU)
The brain of a computer; reads program instructions, processes the data, and carries out programmed operations. (9)

centrifugal pump
A non-positive displacement pump that moves a varying amount of fluid with each rotation using an impeller blade. (6)

closed-loop system
A type of control system that allows feedback to affect the output of the robotic system. (2)

collet gripper
An end effector that delivers 360 degrees of clamping contact. Typically used to pick and place cylindrical parts that are uniform in size. (8)

command resolution
The closest distance between a robot's movements. Command resolution is calculated by dividing the travel distance of each joint by the number of control increments. (4)

commutator
Part of a dc motor that switches current flow. (5)

comparator
A control system element that compares the feedback signal from the controlled element to a reference signal or standard. The comparator develops a correction signal, if needed, which is sent to the control unit. (5)

compiler
A computer program that translates high-level programming languages into machine code. (3)

compliance
An end effector's ability to tolerate the misalignment of mating parts. (8)

compound-wound dc motor
A type of dc motor that has two sets of field windings. (5)

computer vision sensor
A type of vision sensor that detects spatial relationships and provides depth information using stadimetry and triangulation. (7)

conditioning
The process of removing contaminants, such as dirt and moisture, from hydraulic fluid and air. (6)

continuous-path (CP) motion
An extension of point-to-point motion that can involve several thousand points, which creates smooth and continuous movement of the end effector. (3)

control
Devices that alter the flow of power and cause some type of operational change in the system, such as changes in electric current, hydraulic pressure, light intensity, or air flow. (5)

control unit
A component of a microprocessor that receives decoded instructions from the decoder and initiates the proper action. (9)

controller
Part of a robot that coordinates all movements of the mechanical system and receives input from the immediate environment through various sensors. (2)

cost savings
An investment made to replace existing equipment or assets in order to reduce current costs. (12)

counter
A logic device that counts electronic pulses and is used in a variety of applications. (9)

counter electromotive force (cemf)
Voltage generated as the armature of a dc motor rotates. This voltage opposes the voltage applied to the motor. (5)

cycle
Each repeated pattern of change in the direction of the flow of electrons in alternating current power. (5)

cycle timing system
A type of timing system that turns a device on or off at specified intervals or in time with an operational sequence. (5)

cylindrical configurations
A type of work envelope layout which consists of two orthogonal slides, placed at a 90° angle, mounted on a rotary axis. (2)

cylindrical grip
Prehensile movement of the hand in which the fingers and thumb form a "C" shape to grasp a cylindrical object, such as a drinking glass or water bottle. (8)

D

data register
A register compartment that stores information for ALU input. It may also hold an instruction while that instruction is being decoded, or hold data prior to storage in memory. (9)

dc stepping motor
A type of dc motor in which the rotor is a permanent magnet. These motors are primarily used to change electrical pulses into rotary motion. (5)

decade counter
A counter used to change binary signals into a binary-coded decimal (BCD) form. (9)

decoder unit
A component of a microprocessor that examines coded instructions and decides which operation is to be performed by the ALU. (9)

dedicated equipment
Automated equipment that is designed to perform only one function. (4, 12)

degrees of freedom
A value that describes a robot's freedom of motion in three dimensional space—specifically, the ability to move forward and backward, up and down, and to the left and to the right. (2)

delay timing system
A type of timing system that provides a lapse in time before the load device actually becomes energized. (5)

desiccant
A very dry substance used in filters which is designed to attract moisture. (6)

design for manufacturability
The process of designing products with the robots that will assemble them in mind. (4)

detector
A component of that sensing system responds to energy from the source and outputs a signal that is used to control the load device. (5)

detents
An overload protection device comprised of two or more elements held in position by spring-loaded mechanisms. When placed under excessive stress, detents shift from their original position. (5)

digital electronics
The technology that controls robotic and other automated systems. (9)

digital information
Format of computer-processed information that occurs in separate full units; information occurs as "1" and "0" units. (9)

digital input port
The connection through which a robot receives information in the form of digital data. This data is used by the robots computer program to initiate some action. (10)

digital output port
The connection through which the robot controller sends digital data to peripheral equipment. (10)

digital system
An automatic system that processes digital information. Numeric instructions are supplied to the system by magnetic tape, variations in pressure, temperature, or electric current. The numeric signals are decoded and directed to specific machines or machine parts, which then perform the necessary operations. (5)

direct current (dc)
A source of electrical power in which electrons flow in only one direction. (5)

direct-drive motor
A high-torque motor that drives the robot's arm directly, without the need for reducer gears. (2)

direction control device
Devices used to start, stop, or reverse fluid flow without causing a significant change in pressure or flow rate. (6)

dynamic performance
A value that represents how fast the robot can accelerate, decelerate, and come to a stop at a given point. Also called *operational speed*. (4)

dynamic RAM (DRAM)
A RAM circuit that stores bits of data using separate capacitors. To retain the data, the electrical charge of the capacitors must be constantly refreshed. (9)

E

eddy current proximity sensor
A type of proximity sensor that produces a magnetic field in a detector unit. The magnetic field induces eddy currents into any conductive material that is near the probe. A pick-up coil senses a change in magnetic field intensity when an object enters the field. (7)

electric motor
Component of an electrical system that converts electrical energy to mechanical energy in the form of rotary motion. (5)

electromagnetic spectrum
The range of visible and invisible light, which includes frequencies for radio, television, radar, infrared radiation, visible light, ultraviolet light, X-rays, and gamma rays. (7)

electromechanical gripper
An end effector that uses a magnetic field to pick up an object. Also called *magnetic gripper*. (8)

electromechanical system
A type of automated system in which power is transferred from one point to another through mechanical motion that is used to do work. (5)

electronically erasable programmable read-only memory (EEPROM)
Type of ROM that can be erased one bit at a time, using a high electrical charge, and reprogrammed. (9)

end effector
A device that is attached to the end of the manipulator; operates like the robot's hand. Also called the *end-of-arm tooling*. (2)

end stop
End stops are devices placed on an axis to physically block movement past a certain point along the axis. (3)

engineering economics
The analysis of the economic impact and technical aspects of engineering projects. (12)

erasable programmable read-only memory (EPROM)
Type of ROM that can be completely erased, by exposing it to an ultraviolet light, and reprogrammed. (9)

error detector
A component of servo systems that receives data from both the input source and the output device and compares that data to determine if a correction is needed. (5)

error signal
A signal generated when the programmed actions of a robot and its actual actions do not match. (2)

execute
An initiate operation that is based on the instruction that is to be performed. (9)

expandable gripper
An end effector that clamps irregularly shaped workpieces using mechanical fingers equipped with hollow rubber envelopes that expand when pressurized. (8)

expert systems
A system that contains a knowledge base, an inference engine, and a computer-human interface. The information is retrieved as data by users of the system, or is carried out as logic operations. (13)

F

feedback
A signal or data that provides information about the interaction between the control unit and the controlled element. (5)

fetch
A program operation that retrieves the next instruction from memory. (9)

field winding
The coil wrapped around the electromagnet inside a dc motor. (5)

filter
A conditioning device that removes very small pieces of debris and is typically made of some porous medium, such as paper, felt, or very fine wire mesh. (6)

firmware
Programs that are held in a read-only memory. (9)

fixed-sequence robot
A manipulator that repetitively performs successive steps of a given operation according to a predetermined sequence, condition, and position. Its instructions cannot be easily changed. (1)

fixturing
A robot's ability to provide precise points in the process of accurately locating parts. (4)

flexible automation
Type of automation system that includes machines capable of performing a variety of tasks. (1)

flip-flop
A memory device used in digital circuits that can hold an output state even when the input is completely removed, and can change their output based on an appropriate input signal. (9)

flow control device
A control device that alters the volume or flow rate of the fluid. (6)

flow indicator
A device used to test flow rates from pumps and at the inlet and outlet ports of actuators. (6)

fluid motor
A type of motor that converts the force of a moving fluid into rotary motion through the use of vanes, gears, or pistons. (6)

fluid power system
A power system uses air or liquid, or a combination of both, to transfer power. (6)

flux
Strength of the electromagnetic field within a dc motor; directly affected by changes in the field current. (5)

force
Any factor that tends to produce or modify the motion of an object; normally expressed in units of weight. (6)

FRL unit
A conditioning device used in many pneumatic systems that combines the air filter, regulator, and lubricator components. (6)

G

general-purpose interface bus (GPIB)
A cable that interfaces system controllers with programmable instruments, and can support a maximum of 15 devices. (10)

gripper
An end effector that performs prehensile movements by grasping objects and moving them. (8)

H

hard automation
Type of automation system that uses machinery specifically designed and built to perform particular tasks within an assembly line. (1)

heat exchanger
A device that cools the fluid in a hydraulic system to maintain a constant fluid temperature. (6)

hexadecimal number system
A base 16 number system that is used to process large numbers. The largest number used in a place is 15, with digits 0-9 and letters A-F. (9)

hierarchical control
A system of control in which components are organized into levels. Each level is dependent on the level above for instructions. (2)

hierarchical control programming
A system of programming in which each program level accepts commands from the level above and responds by generating simplified commands for the level below. (3)

high-level language
Programming languages that more closely resemble standard English than traditional programming code. (3)

hook movement
Nonprehensile hand movement that involves curling the tips of the fingers to pull or lift objects. (8)

hydraulic drive
A power system that uses fluid and consists of a pump connected to a reservoir tank, control valves, and a hydraulic actuator. (2)

I

I/O port
Input/output through which digital signals used for communication travel from system to system. (10)

image acquisition
A function of machine vision in which a workpiece is illuminated and an image is digitally scanned. (10)

image analysis
A function of machine vision in which information from the image is gathered and analyzed by computer software. Using algorithms, the software identifies and measures features of the digital image. (10)

image interpretation
A function of machine vision in which image information is interpreted and the robot makes decisions about tasks that must be performed. (10)

image preprocessing
A function of machine vision in which an analog-to-digital converter changes the analog signal of the image into an equivalent digital signal. The digital signal represents light intensity values over the entire image. (10)

indicator
A subsystem that displays information about operating conditions at various points throughout the system. (5)

inductance
A property of electrical circuits caused by the magnetic field that surrounds a coil when current is flowing. In ac circuits, inductance opposes changes in current and increases as frequency increases. (7)

inductive transducer
A type of transducer with a stationary coil and a movable core. An inductive transducer measures movement and creates signals that affect current flow. (7)

industrial robot
A multifunction manipulator programmed to perform various tasks. (1)

inertia
The resistance to change of an object to be moved. (6)

infrared sensor
A type of light sensor that responds to radiation in the infrared region of the electromagnetic spectrum. (7)

input-output (I/O) transfer
Basic computer system function in which an I/O port actuates to either receive data from an input or send data to an output device. (9)

intelligent robot
A robot that can, by itself, detect changes in the work environment by means of sensors (visual and/or tactile). Using its decision-making capability, it can then proceed with the appropriate operations. (1)

interface
The common point at which two or more systems communicate with each other. (10)

interlocks
Safety devices that are designed to prevent unauthorized access to hazardous areas by requiring a key for entry. (4)

interrupt
A signal that originates from peripheral equipment to inform the MPU that the device needs attention. (9)

interval timing system
A type of timing system that is used after a load is energized and operates using specified time intervals. (5)

inverter
A type of logic gate that has one input and one output. The output of an inverter is opposite to the input. Also called a *NOT gate*. (9)

L

laser
Acronym for light amplification by stimulated emission of radiation. A highly focused beam of light that can cut, carry messages, and perform other kinds of work. (7)

laser interferometric gauge
A range sensing system that is sensitive to humidity, temperature, and vibration. (7)

lateral grip
Prehensile movement of the hand in which the fingers and thumb grasp flat objects from the sides, rather than around. (8)

light curtain
A barrier which consists of photoelectric, presence-sensing devices. The sensors send a signal to cut the power to the robot if a worker enters the area. (4)

light-emitting diode (LED)
Small, lightweight opto-electronic device; solid-state lamp. The semiconductors used in LEDs produce light when an electric current is applied. (7)

light pipes
Fiber-optic rods used in a transmission path to direct light energy around corners or to unusual locations. (5)

limit switch
An electrical switch with an actuator that is mechanically linked to a set of contacts. When activated, contacts that are normally open are closed and contacts that are normally closed are opened. This causes a change of state to an electrical circuit. (7)

linear actuator
A type of actuator that provides motion along a straight line. (2)

load
The part or parts of an automated system designed to produce work. (5)

logic circuit
The electronic circuitry used with robots that provides commands to control motors, hydraulic systems, and pneumatic systems. These circuits also store information, count, encode, and decode. (9)

logic gate
Basic binary logic circuits pass or block certain signals (AND circuit, OR circuit, and NOT circuit). (9)

lubricator
A conditioning device used in pneumatic power systems that adds a small quantity of oil to the air after it leaves the regulator. (6)

M

machine vision system
Video systems that are built into automated work cells and are used for applications such as guidance, part orientation, measuring and inspecting parts, and image identification. (10)

magnetic field sensor
A type of sensor that identifies a change in an existing magnetic field without making physical contact with objects in the environment. (7)

magnetic gripper
An end effector that uses a magnetic field to pick up an object. Also called *electromechanical gripper*. (8)

manipulator
The arm of the robot which must move materials, parts, tools, or special devices through various motions to provide useful work. (2)

manual manipulator
A manipulator worked by a human operator. (1)

manual programming
A type of machine setup in which an operator adjusts the necessary end stops, switches, cams, electric wires, or hoses to set up the robot's movement sequence. (3)

manual rate control box
A control device that consists of a knob and some switches used to move each axis of a robotic arm in the process of programming a path of motion. (3)

mechanical finger gripper
An end effector used for grasping parts within a confined space, reaching into channels, or picking and placing any object that has a simple shape. Typically arranged in two-, three-, and four-finger configurations. (8)

mechanical fuses
Overload protection devices in the form of pins or tubes that break or buckle under extreme stress. (5)

memory
The ability of a MPU to store data, so a single bit or a group of bits can be easily retrieved. (9)

metering
Controlling the rate of fluid flow. (6)

microbots
Miniature robots that can measure less than a millimeter in size. (13)

microprocessor unit (MPU)
A microchip that combines the arithmetic logic unit and control circuitry of the computer. (9)

microswitch
An electrical switch that is turned on or off with a very small amount of force. Microswitches are small, durable, and easy to activate. (7)

MPU cycle
The timeframe in which system operations are performed. (9)

N

NAND gate
A NOT gate combined with an AND gate. (9)

nanometer (nm)
A unit of light measurement. A nanometer is one billionth of a meter. (7)

non-positive displacement pump
A type of fluid pump that moves fluid with each rotation of the impeller blades. The amount of fluid that passes through the pump with each rotation varies. (6)

non-servo robot
The simplest type of robots; uses an open-loop system that does *not* allow for system feedback. Often referred to as *limited sequence*, *pick-and-place*, or *fixed-stop robots*. (2)

Glossary

nonprehensile movement
Movements of an end effector that do not require particular finger dexterity or use of the opposed thumb. Examples include pushing, poking, punching, and hooking. (8)

NOR gate
A NOT gate combined with an OR gate. (9)

NOT gate
A type of logic gate that has one input and one output. The output of a NOT gate is opposite to the input. Also called an *inverter*. (9)

numerically controlled (NC) robot
A manipulator that can perform the sequence of movement, conditions, and positions of a given task which are communicated by means of numerical data. (1)

O

octal number system
A base 8 number system used to process large numbers. The place values of digits, moving to the left of the octal point, are powers of eight. (9)

off-line programming
Creating the programming for a robot on a computer that is not connected to the robot. (3)

on-line programming
Programming a robot using a computer at the robot's console. (3)

open-loop system
A type of control system that does *not* include a feedback mechanism to compare programmed positions to actual positions. (2)

operational speed
A value that represents how fast the robot can accelerate, decelerate, and come to a stop at a given point. Also called *dynamic performance*. (4)

oppositional grip
Prehensile movement of the hand in which the tip of the index finger and thumb hold an object. (8)

optical fibers
Fiber made of glass or plastic that can transmit light from one point to another regardless of how the material is bent or shaped. (7)

optical proximity sensor
Type of proximity sensor that measures the amount of light reflected from an object. An optical proximity sensor can respond to either visible or infrared light. (7)

opto-electronic
Describes devices that use a combination of optical and electronic components. (7)

OR gate
A type of logic circuit that has two or more inputs and one output. (9)

overload sensor
A sensor used on an end effector that detects obstructions or overload conditions within fractions of a second. (8)

P

palmar grip
Prehensile movement of the hand in which the fingers and thumb wrap around an object, such as a baby's rattle or a human finger, to grasp. (8)

parallel port
Type of interface connection through which the robot controller sends digital data to peripheral equipment. (10)

parallel transmission
An interface method in which multiple bits of data are sent at the same time, following side-by-side paths, like the lanes of a highway. (10)

Pascal's law
A principle of fluid properties that states when pressure is applied to a confined fluid, the pressure is transmitted, undiminished, throughout the fluid. Additionally, this pressure acts on all surfaces of the container, at right angles to those surfaces. Discovered by Blaise Pascal (1653). (6)

payback period
The time required to recover the money spent for new equipment, either through savings in labor or materials. (12)

payload
The maximum weight or mass of material a robot is capable of handling on a continuous basis. (4)

period
The time interval for a pulse to pass through a complete MPU cycle from beginning to end. (9)

permanent-magnet dc motor
A type of dc motor in which the dc power supply is connected directly to the conductors of the rotor through the brush-commutator assembly. This type of motor is used when a low amount of torque is needed. (5)

personal computer (PC)
Small desktop or portable computing units built around a single integrated circuit. (9)

photoconductive device
Type of opto-electronic device that varies in conductivity according to fluctuations in light. Electrical resistance decreases when light is more intense and increases when light intensity decreases. (7)

photoemissive device
Type of opto-electronic device that emits electrons in the presence of light. (7)

photovoltaic device
Type of opto-electronic device that converts light energy into electrical energy. When light energy falls on a photovoltaic device, it creates an electrical voltage. Also called *solar cell*. (7)

pick-and-place motion
Limited-sequence robots use pick-and-place motion to move the end effector to the correct position. Pick-and-place motion is often used in manufacturing processes to perform work that is repetitive and does not require many complicated movements to accomplish a task. (3)

piezoelectric effect
The characteristic of certain crystals to develop an electrical potential when subjected to mechanical stress. Rochelle salt and quartz are examples of these crystals. (7)

pitch
The up-and-down movement of a robot's wrist. (2)

pixel
Abbreviation for picture elements. Term used for the dots of light that form an image. (10)

place value
The position of a digit in a number system with respect to the decimal point. (9)

playback robot
A manipulator that can reproduce operations originally executed under human control. An operator initially feeds in the instructions relating to sequence of movement, conditions, and positions, and the instructions are then stored in memory. (1)

pneumatic drive
A power system that uses air-driven actuators. (2)

point-to-point (PTP) motion
The movement of a robot's end effector through a number of points in space to position at a desired location. (3)

positive displacement pump
A type of fluid pump that moves a definite amount of fluid through the pump during each revolution. (6)

positive logic
Binary systems that use the digits 1 and 0 to represent two operational conditions. (9)

power
A measurement that considers the amount of work accomplished in relation to the amount of time taken to perform the work. (6)

power supply
A mechanism that provides energy to drive the controller and actuators. (2)

prehensile movement
Movement of the hand that requires the use of the thumb, in addition to the fingers, to grasp objects. (8)

preloaded springs
Overload protection devices that respond to excess stress by releasing a spring that causes the end effector to break away from the work area. (5)

pressure
The amount of force applied to a specific area; often expressed in pounds per square inch (lb./in^2 or psi) or as newtons per square meter (N/m^2). (6)

pressure indicator
A device that monitors the fluid pressure within a hydraulic system. (6)

pressure regulator valve
A type of valve that allows the air pressure in a pneumatic system to be adjusted to a specific level. (6)

pressure-relief valve
A type of valve that controls the pressure in a system. In a hydraulic system, the pressure relief valve allows fluid to flow back into the reservoir when the pressure rises to a dangerous level. In a pneumatic system, pressure-relief valves release excess air in the system into the atmosphere. (6)

pressure sensitive safety mat
A sensing system that contains sensors embedded in a floor mat. A signal is sent to the controller when weight is placed anywhere on the mat. (4)

preventative maintenance
A scheduled plan of maintenance tasks that are performed to prevent equipment breakdown and loss of production. (11)

prime mover
The component of a power system that provides the initial power for movement in the system. (6)

program
A series of instructions stored in the controller's memory. (2)

program counter
A memory device that indicates the location in memory of either the instruction currently being performed or the next instruction to be executed. (9)

programmable read-only memory (PROM)
Type of ROM chip that has data electrically burned onto it and cannot be reprogrammed. (9)

proximity sensor
Type of sensor that detects either the absence of an object or the presence of an object within a certain distance. (7)

R

radial traverse
One of the three degrees of freedom in the robot arm; involves the extension and retraction of a robot's arm, which creates in-and-out motion relative to the base. (2)

random access memory (RAM)
Type of computer system memory that can be altered; information may be retrieved from *and* written to random access memory. (9)

range sensor
Type of sensor that can determine the precise distance from the sensor to an object. (7)

read-only memory (ROM)
Type of computer system memory that is permanent and cannot be changed. Data recorded in read-only memory often includes operating system and program information. (9)

reciprocating pump
A type of positive displacement pump that uses the reciprocating action of a moving piston to move fluid into and out of a chamber. (6)

rectification
The process of converting alternating current (ac) to direct current (dc) power. (5)

reed switch
A simple type of magnetic field sensors that either makes or breaks contact in response to changes in a magnetic field. (7)

register unit
A storage component of a microprocessor consisting of several compartments that store information. Each compartment in a register unit stores a specific type of data. (9)

remote-center compliance (RCC) device
A device that is installed in the wrist of the robot to help compensate for workpiece misalignment or irregularities. (8)

repeatability
A value that expresses how close the robot's hand will actually come to the designated position while repetitively performing the task or procedure. (4)

reprogrammable
A robot's ability to be given new instructions that meet changed requirements and perform new tasks. (1)

resin-core solder
A type of solder that is noncorrosive and nonconductive. Resin-core solder is always used on electronic circuits. (11)

resistance
Friction that forms in a fluid system and causes a decrease of power from the input to the output. (6)

resistive transducer
Type of transducer that converts a variation in resistance into electrical variations. (7)

resolution
The smallest incremental movement a robot can make, and is determined by the robot's control system. (4)

revolute configuration
A type of work envelope layout that is irregularly shaped and requires a jointed-arm robot. (2)

rework
The process of fixing parts that no longer meet product specifications. (11)

robot
A machine or device that automatically performs tasks or activities which are typically completed using *human* skill and intelligence. (1)

robotics
The study, engineering, and use of robots. (1)

roll
The rotation or swivel movement of a robot's wrist. (2)

rotary actuator
A type of actuator that provides rotation, moving a load in an arc or circle. (2)

rotary gear pump
A type of positive displacement pump that uses rotary motion to produce pumping action. (6)

rotary vane pump
A type of positive displacement pump that uses a series of sliding vanes to move fluids. (6)

rotational traverse
One of the three degrees of freedom in the robot arm; involves movement on a vertical axis. This is the side-to-side swivel of the robot's arm on its base. (2)

rotor
The rotating component of a motor that includes the armature, shaft, and associated parts. (5)

S

SCARA
Selective compliance assembly robot arm. A robot arm configuration that is horizontally articulated and generally has one vertical (linear) and two revolute joints. (2)

scrap
A manufactured part that cannot be fixed and must be discarded. (11)

sensing system
An electrical system that signals a response to a particular form of energy, such as light. (5)

sensory feedback
Input received through various system sensors about the environment. (3)

serbots
An abbreviation for *service robot*. (13)

serial port
Type of interface connection through which a computer sends or receives digital data using serial transmission. (10)

serial transmission
An interface method in which data is transmitted one bit at a time; used for delivering digital data over long distances. (10)

series-wound dc motor
A type of dc motor in which the armature (rotor) and field circuits are connected in a series arrangement. This is the only type of dc motor that also can be operated using ac power. Also called *universal motors*. (5)

service robot
A mobile robot that can move to the work area to perform the necessary tasks. (4)

servo amplifier
A system component that translates feedback signals from the controller into motor voltage and current signals. (2)

servo robot
A robot system that uses a closed-loop system. Servo robot systems allow feedback signals to affect the robot's performance. (2)

servo system
Machines that change the position or speed of a mechanical object in response to system feedback or error signals. (5)

servomotor
A component of a servo system that produces controlled shaft displacements to achieve a precise degree of rotary motion. (5)

shunt-wound dc motor
A type of dc motor in which the field coils are connected in parallel with the armature (rotor) and have relatively high resistance. (5)

single-phase ac motor
A type of ac motor that operates using a single-phase ac power source. (5)

single-phase induction motor
A type of ac motor that has a solid rotor and must be set into motion by some auxiliary starting method. (5)

slip
The difference between the synchronous speed and the rotor speed of a motor. (5)

software
Programs that are transferred to a computer system by keyboard, CD-ROM, or download, and the instructions are stored in read/write memory. (9)

solar cell
Type of opto-electronic device that converts light energy into electrical energy. When light energy falls on a solar cell, it creates an electrical voltage. Also called *photovoltaic device*. (7)

solder sucker
A device that creates suction to remove unwanted solder from a printed circuit board without damaging the board. (11)

sound sensor
Type of sensor that relies on the piezoelectric effect to convert sound to electrical energy. (7)

spatial resolution
A value that expresses the accuracy of movement of the robot's tool tip. Spatial resolution takes into account command resolution and mechanical inaccuracy. (4)

speed sensing
Type of sensor that measures the rotary motion of shafts, gears, pulleys, and other rotating components of industrial equipment. The voltage output or light reflected is translated into speed readings. (7)

spherical configuration
A type of work envelope layout that resembles a sphere, relative to the base. Also called *polar configuration*. (2)

spherical grip
A prehensile movement of the hand in which the fingers are used to hold round objects, such as holding and throwing a baseball. (8)

spread movement
A nonprehensile movement of the hand in which the fingers and thumb are extended outward until they make contact with the interior walls of a hollow object. The force of the fingers against the walls of the object allows it to be picked up and carried. (8)

squirrel cage rotor
The solid rotor in a single-phase induction motor. (5)

stadimetry
A factor to determine depth that considers the distance to an object based on the apparent size of the camera image. (7)

static RAM (SRAM)
A RAM circuit that uses semiconductor devices, called flip-flops, to store data. (9)

stator
The stationary portion of a motor that includes the permanent magnets, a frame, and other stationary components. (5)

strain gauge
A sensor made of fine-gage resistance wire mounted on a strip of insulation. A strain gauge measures mechanical movement. (7)

strainer
An in-line device that captures large particles of foreign matter within hydraulic systems. (6)

subroutine
A set of instructions within programming code that has an independent beginning and end. (3)

subsystem
Various components that work together to form a unit. Subsystems in a robotic system may include several different types of inputs, controls, and specialized machinery for proper operation. (5)

synchronous motor
A type of servomotor that is comprised of a rotor and a stator assembly; contains no brushes, commutators, or slip rings. (5)

synchronous speed
Speed of the rotating magnetic field of an ac motor. (5)

synthesized system
The combination of subsystems into a cohesive, operational process or system. (5)

system
A combination of components that work together to form a unit. (5)

T

tachometer
A device used to measure the speed of an object. In the case of robotic systems, a tachometer is used to monitor acceleration and deceleration of the manipulator's movements. (2)

tactile sensor
Type of sensor that indicates the presence of an object by touch. (7)

task-level programming
A method of robot programming in which the goals of each task, rather than the motions or points, are entered using simple English-like terms. (3)

teach pendant
A device used to teach a robot the movements required to perform a useful task. Also called a *teach box* or *hand-held programmer*. (2)

teach pendant programming
A method of robot programming in which the operator leads the robot through the various positions involved in an operation. Each desired point is recorded by pushing buttons on the teach pendant. The recorded points are used to generate a point-to-point path the robot follows during operation. (2)

telerobotics
Field of robotics research and development that focuses on robots operated by remote control. (13)

thermistor
A temperature-sensitive resistor commonly used for heat sensing. The resistance of a thermistor decreases as temperature increases (and vice-versa). (7)

thermocouple
Device that converts heat energy into electrical energy. (7)

thermoelectric sensor
Type of sensor that produces a change in electrical output in response to fluctuations in temperature. (7)

three-phase ac motor
A type of motor that uses a three-phase ac power source. (5)

three-phase induction motor
A type of motor that uses a three-phase ac power source and has a squirrel cage rotor. (5)

three-phase synchronous motor
A type of motor in which direct current is applied to the wound rotor to produce an electromagnetic field and three-phase ac power is applied to the stator. (5)

timing system
An electrical system that turns a device on or off at a specific time or in step with an operating sequence. (5)

tool
An end effector that executes nonprehensile movements to perform specific tasks, such as welding or painting. (8)

torque
The rotary motion, or turning force, that is produced by the armature. (5)

touch-sensitive proximity sensor
Type of proximity sensor that operates on capacitance developed by a large conductive object (such as the human body). This capacitance changes the frequency of an electronic circuit. A conductive plate or rod may be used to sense contact. (7)

trajectory
Planned path of movement. (2)

transducer
Part of a sensor that converts light, heat, or mechanical energy into electrical energy. (7)

transmission path
A subsystem that provides a channel for the transfer of energy. (5)

triangulation
The process of measuring angles and the base line of a triangle to determine the position of an object. (7)

troubleshooting
The systematic process of identifying and correcting problems in an operation or system. (11)

truth table
In computer circuitry, a table that displays combinations of inputs and resulting outputs of a logic gate. (9)

turbulence
Describes how fluid moves through a fluid power system. Conditions such as size and smoothness of internal surfaces, temperature of the fluid, and the location and number of valves and fittings may cause irregular flow characteristics. (6)

U

ultraviolet sensor
Type of sensor that responds to electromagnetic radiation in the ultraviolet. (7)

Unimate
The first robot produced by Unimation (1961). (1)

universal motor
A motor that can be powered by either an ac or dc source and is built like a series-wound dc motor. (5)

V

vacuum gripper
An end effector that has one or more suction cups made of natural or synthetic rubber. The suction cups attach to the surface of a workpiece to grip. (8)

variable-sequence robot
A manipulator that is similar to the fixed-sequence robot, but its sequence of movement can be easily changed. (1)

vertical traverse
One of the three degrees of freedom in the robot arm. This is the up-and-down motion of the robot's arm. (2)

voice recognition
An input system in which recognizable words or phrases are used as a form of audio data entry, rather than using a keyboard. Also called *speech recognition*. (3)

weight
The gravitational force exerted on a body by the earth; expressed in pounds (lb.) in the U.S. customary measurement system and in newtons (N) in the metric system. (6)

work
A measurement of what a system actually accomplishes. Work occurs when energy is transformed into mechanical motion, heat, light, chemical action, or sound. (5)

work envelope
The area within reach of a robot's end effector. (2)

W

walk-through programming
A programming method used for continuous-path robots in which an operator physically moves the end effector through the desired motions. As the robot moves along the desired path, as many as several thousand points are recorded into memory. (3)

WAVE
The first robot programming language; developed at the Stanford Artificial Intelligence Laboratory in 1973. (3)

X

X-rays
An invisible band of radiation within the electromagnetic spectrum between ultraviolet and gamma rays. (7)

Y

yaw
The side-to-side movement of a robot's wrist. (2)

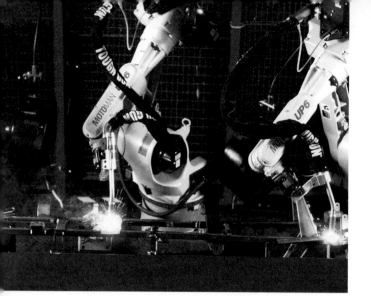

Index

A

ac motors, 129–132
accumulator, 215
accuracy, 91
acoustical proximity sensors, 191
actuator control, 27
actuator drive, 41–46
actuators, 26
 linear, 28
 rotary, 28
address register, 215
alternating current (ac), 117
analog information, 212
AND gate, 227–228
angstrom (Å), 176
arithmetic logic unit (ALU), 215
armature, 123
articulated configuration, 47, 50
artificial intelligence (AI), 14, 61, 285–287
assembly, 104–107
automated guided vehicles (AGVs), 108–109
automated systems and subsystems, 116–118
automatic tool changers, 204
automation, types, 18–21
automaton, 12
avoidance costs, 274

B

bifilar construction, 136
binary-coded-decimal (BCD) number system, 222
binary counters, 233–235
binary logic circuits, 226–236
 digital counters, 232–236
 flip-flops, 230–232
 logic gates, 227–230
binary number system, 220–222
binary point, 220
bistable device, 227
bit, 212
brushes, 124
buses, 215
bus network, 215
byte, 212

C

capacitance, 172
capacitive transducers, 172, 175
capital investment, 273
careers, 289
 field service technician, 255
 robotics engineer, 109
 software engineer, 60

Cartesian configuration, 51
central processing unit (CPU), 213
centrifugal pumps, 153–154
closed-loop system, 39
collet grippers, 201
combination logic gates, 229–230
command resolution, 89–90
commutator, 124
comparator, 122
compiler, 72
compliance, 206
compound-wound dc motor, 128
computer, 14
computer systems, 212–219
    components, 214–217
        basic functions, 217–219
        programming, 236–239
computer vision sensors, 186
conditioning, 143
    fluid conditioning devices, 154–156
configurations, 47–56
continuous-path (CP) motion, 66–67
control, 118
control devices, 158–161
controller, 26–27
control systems, 121–123
    types, 36–41
control unit, 215
cost savings, 274
counter, 232–233
counter electromotive force (cemf), 124–126
CyberKnife® System, 208
cycle, 117
cycle timing system, 121
cylindrical configuration, 53–55
cylindrical grip, 196

D

data register, 215
dc motors, 124–128
dc stepping motors, 135–136
decade counters, 235–236
decoder unit, 215
degrees of freedom (DOF), 30, 32–34, 89
delay timing systems, 121
desiccant, 155
design for manufacturability, 82–83
detector, 120
detents, 137
Devol, George C., 14, 16
die casting, 95–96
digital control systems, 123
digital counters, 232–236
digital electronics, 212
digital information, 212
digital input ports, 242–243
digital number systems, 219–226
    binary, 220–222
    hexadecimal, 225–226
    octal, 222–225
digital output ports, 243
direct current (dc), 117
direct-drive motors, 44
direction control, 158–160
    devices, 158
dynamic performance, 92
dynamic RAM (DRAM), 217

E

early robots, 12–14
eddy current proximity sensors, 191
electrical systems, 119–137
    rotary motion systems, 123–137
        control systems, 121–123
        sensing systems, 120–121
        timing systems, 121
electrically erasable programmable
    read-only memory (EEPROM), 216
electric drive, 42, 44
electric motors, 123

electromagnetic spectrum, 176
electromechanical grippers, 202–203
electromechanical systems, 115–116
end effectors, 29–30, 195
    changeable, 204
    design, 205–207
    movement, 196
    types, 198–204
energy source, 117
engineering economics, 274
erasable programmable read-only memory (EPROM), 216
error detector, 133
error signal, 40
execute, 218
expandable grippers, 206
expert systems, 285–287

F

feedback, 122
fetch, 218
fiber optics, 180
field service technician, 255
field windings, 124
filter, 155
firmware, 236
first-generation robots, 60–61
fixed-sequence robot, 18
fixturing, 102
flexible automation, 20–21
flip-flops, 230–232
flow control devices, 161
flow indicators, 167
fluid conditioning devices, 154–156
fluid flow characteristics, 144–146
fluid motors, 166–167
fluid power principles, 146–148
fluid power systems, 139–140
    models, 140–144
fluid power system components, 149–167
    control devices, 158–161

fluid pumps, 149–154
indicators, 167
load devices, 163–167
transmission lines, 158
fluid pumps, 149–154
    centrifugal, 153–154
    reciprocating, 149–150
    rotary gear, 150, 152
    rotary vane, 152–153
flux, 128
force, 146
four-finger grippers, 201
FRL unit, 156
fully-automated factories, 282

G

gas lasers, 182–183
general-purpose interface bus (GPIB), 246–247
generations, 60–61
generators, 18
grippers, 198–203
    collet grippers, 201
    electromechanical grippers, 202–203
    mechanical finger grippers, 198–201
    vacuum grippers, 202

H

hard automation, 18–19
heat exchanger, 155
hexadecimal number system, 225–226
hierarchical control, 26–27
    programming, 73–75
high-level languages, 72
Holonyak, Nick Jr., 193
hook movement, 196
hybrid systems, 167
hydraulic conditioning, 155
hydraulic drive, 44–45
hydraulic system model, 140–142

I

I/O port, 242
image acquisition, 249
image analysis, 250
image interpretation, 250
image preprocessing, 249
implementation plan, 275–279
    comparison matrix, 276
indicators, 118, 167
inductance, 175
inductive transducers, 175
industrial robots, 14–18
    anthropomorphic, 15
    early, 14–16
    evolution, 14–16, 19
industrial use of robots, 93–112
    assembly, 104–107
    die casting, 95–96
    inspection, 107–108
    machine loading and unloading, 94
    machining processes, 102–104
    material handling, 108–109
    pick-and-place, 93–94
    service, 109–111
    spraying operations, 99–100
    welding, 96–98
inertia, 146
infrared sensors, 179
input-output (I/O) transfer, 218
inspection, 107–108
intelligent robot, 18
interfacing, 242–247
    digital input ports, 242–243
    digital output ports, 243
    parallel ports, 246–247
    serial ports, 245–246
interlocks, 85
interrupt, 218–219
interval timing systems, 121
inverter, 228
investment pay off, 268

L

laser, 181
laser interferometric gauge, 192
lateral grip, 196
light curtain, 87
light-emitting diode (LED), 190, 193
light pipes, 120
light sensors, 176–184
    fiber optics, 180
    gas lasers, 182–183
    infrared, 179
    laser, 181
    semiconductor lasers, 183
    ultraviolet, 179
    X-rays, 184
limit switches, 192
linear actuators, 28, 163–165
linear variable differential transformer
    (LVDT), 175
literature, 13–14
load, 118
load capacity, 92
load devices, 163–167
logic circuit, 212
logic gates, 227–230
lubricator, 156

M

machine loading and unloading, 94
machine vision, 247–250
machining and assembly tools, 204
machining processes, 102–104
magnetic field sensors, 188
magnetic grippers, 202–203
main control, 27
maintenance programs, 279
    development, 262–263
    implementation, 263
manipulator, 27–29
manual manipulator, 18

manual programming, 67–68
manual rate control box, 74
manufacturing, 82–88
    robot design, 82–83
        robot productivity, 266
        robot safety considerations, 85–88
        robots used in, 265–279
material application tools, 204
material handling, 108–109
mechanical finger grippers, 198–201
mechanical fuses, 137
mechanical movement sensors, 190
mechanical systems, 119
memory, 215–217
metering, 161
microbots, 284
microprocessor unit (MPU), 214–215
microswitch, 192
motion control, 63–67
    continuous-path, 66–67
    pick-and-place, 63–64
    point-to-point, 64–66
MPU cycle, 217
    period, 217

N

NAND gate, 229
nanometer (nm), 176
non-industrial uses of robots, 282–284
non-positive displacement pump, 149
nonprehensile movements, 196
non-servo robots, 36–39
NOR gate, 230
NOT gate, 228
numerically controlled (NC) robot, 18

O

octal number system, 222–225
off-line programming, 70
on-line programming, 70

open-loop system, 36
operational speed, 92
oppositional grip, 196
optical fibers, 180
optical proximity sensors, 190
opto-electronic, 176
OR gate, 228
overload protection, 137
overload sensors, 206

P

palmar grip, 196
parallel ports, 246–247
parallel transmission, 246
Pascal's law, 146
path control, 27
payback period, 274
payload, 92
permanent-magnet dc motor, 127
personal computer (PC), 213
photoconductive devices, 176
photoemissive devices, 176
photovoltaic devices, 176
pick-and-place motion, 63–64, 93–94
piezoelectric effect, 186
pitch, 32
pixels, 249
place value, 219
playback robot, 18
pneumatic conditioning, 155–156
pneumatic drive, 45–46
pneumatic system model, 143–144
point-to-point motion, 64–66
polar configuration, 55–56
positive displacement pump, 149
positive logic, 222
potential applications, 276
power, 147
power supply, 30
prehensile movements, 196

preloaded springs, 137
pressure, 147
pressure control, 158
pressure indicators, 142
pressure regulator valve, 143
pressure relief valve, 141
pressure sensitive safety mat, 88
preventative maintenance, 261–263
prime mover, 140
productivity, 266
program, 26
program counter, 215
programmable read-only memory (PROM), 216
programming, 59–79
    computer terminal use, 70
    computer, 236–239
    evolution, 60–61
    hierarchical control programming, 73–75
    languages, 70–72
    manual, 67–68
    methods, 30, 67–70
    task-level programming, 75–77
    teach pendant, 69–70
    walk-through, 70
program planning, 239
proximity sensors, 190–192

R

radial traverse, 32
random access memory (RAM), 217
range sensors, 192
read memory function, 218
read-only memory (ROM), 215–216
reciprocating pumps, 149–150
rectification, 117
reed switch, 188
register unit, 215
remote center compliance (RCC) device, 206
repeatability, 91
reprogrammable robot, 17

resin-core solder, 260
resistance, 144
resistive transducers, 172
resolution, 89–90
revolute configuration, 47, 50
rework, 261
robotics
    definition, 14
    impact on society, 287–289
robotics engineer, 109
robots, 12
    classifying, 34–57
    early, 12–14
    economic justifications, 274–275
    evaluating uses, 269–275
    evolution, 14–16
    implementation, 268
    industrial use. *See* industrial use of robots
    in literature, 13–14
    in manufacturing, 82–93, 265–279
    investment in, 268
    non-economic justifications, 273–274
    non-industrial uses, 282–284
    non-servo, 36–39
    origin of name, 13
    parts, 24–30
    safety considerations, 85–88
    selection, 89–92
    servo, 39–41
roll, 32
rotary actuators, 28, 165
    electric, 132
rotary gear pumps, 150, 152
rotary motion systems, 123–137
    ac motors, 129–132
    dc motors, 124–128
rotary vane pumps, 152–153
rotational traverse, 32
rotor, 124
RQ-4 Global Hawk, 238

S

safety
    considerations, 85–88
    barriers and sensors, 86–88
SCARA, 47
scrap, 261
second-generation robots, 61
selective compliance assembly robot arm, 47
semiconductor lasers, 183
sensing systems, 120–121
sensor applications, 189–192
sensors, 171–192
    function, 172–175
    types, 175–189
sensory feedback, 73
serbot, 282–283
serial ports, 245–246
serial transmission, 245
series-wound dc motor, 127
service robots, 109–111, 282–283
servicing techniques, 256–261
servo amplifier, 39
servomotor, 133
servo robots, 39–41
servo systems, 132–137
shunt-wound dc motor, 128
single-phase ac motors, 129
single-phase induction motor, 129
slip, 131
software, 236
software engineer, 60
solder sucker, 260
sound sensors, 186
spatial resolution, 90
speed sensing, 189
speed sensors, 189–190
spherical configuration, 55–56
spherical grip, 196
spraying operations, 99–100
spread movement, 196
squirrel cage rotor, 129
stadimetry, 186
static RAM (SRAM), 217
stator, 124
strainer, 155
strain gauge, 190
subroutine, 72
subsystems, 116
sweep, 47
synchronous motor, 133
synchronous speed, 131
synthesized system, 116
system, 116

T

tachometer, 29
tactile sensors, 192
task-level programming, 75–77
teach pendant, 30, 69–70
telerobotics, 283
temperature sensors, 187–188
thermistor, 187
thermocouples, 188
thermoelectric sensors, 187
third-generation robots, 61
three-finger grippers, 201
three-phase ac motors, 131
three-phase induction motors, 131
three-phase synchronous motors, 131–132
timing function, 217–218
timing system, 121
tools, 198, 203–204
torque, 124
touch-sensitive proximity sensors, 192
training, 278–279
trajectory, 27
transducers, 172, 175
transmission lines, 158
transmission path, 117
triangulation, 186

troubleshooting, 254–256
truth table, 228
turbulence, 144
two-finger grippers, 198, 201

U

ultraviolet sensors, 179
Unimate, 14–15
universal motor, 129

V

vacuum grippers, 202
variable-sequence robot, 18
vertical traverse, 32
vertically articulated configuration, 47
vision sensors, 184–186
voice recognition, 77, 79

W

walk-through programming, 70
WAVE, 70
weight, 147
welding, 96–99
    tools, 203
work, 118, 147
work envelope, 30, 89
    shape, 46–57
write memory, 218

X

X-rays, 184

Y

yaw, 32